30 Climate COPs Later

30 Climate COPs Later

Stories from Canadian Participants

Edited by Thomas Burelli,
Alexandre Lillo, Lynda Hubert Ta,
Lauren Touchant, and Elie Klee

University of Ottawa Press
2025

Les **Presses** de l'Université d'Ottawa
University of Ottawa **Press**

Les Presses de l'Université d'Ottawa / University of Ottawa Press (PUO-UOP) is North America's flagship bilingual university press, affiliated to one of Canada's top research universities. PUO-UOP enriches the intellectual and cultural discourse of our increasingly knowledge-based and globalized world with peer-reviewed, award-winning books.

www.Press.uOttawa.ca

Library and Archives Canada Cataloguing in Publication

Title: 30 climate COPs later : stories from Canadian participants / edited by Thomas Burelli, Alexandre Lillo, Lynda Hubert Ta, Lauren Touchant, and Elie Klee.
Other titles: Thirty climate COPs later
Names: Burelli, Thomas, editor. | Lillo, Alexandre, editor. | Hubert Ta, Lynda, editor. | Touchant, Lauren, editor. | Klee, Elie, editor.
Description: Includes bibliographical references.
Identifiers: Canadiana (print) 20250238659 | Canadiana (ebook) 20250238713 | ISBN 9780776645575 (softcover) | ISBN 9780776645599 (PDF) | ISBN 9780776645582 (EPUB)
Subjects: LCSH: Conference of the Parties (United Nations Framework Convention on Climate Change) | LCSH: Climatic changes—Government policy—International cooperation. | LCSH: Climatic changes—Government policy—Canada. | LCSH: Climate change mitigation—International cooperation.
Classification: LCC QC903 .A13 2025 | DDC 363.738/74561—dc23

Legal Deposit: Fourth Quarter 2025
Library and Archives Canada

© University of Ottawa Press 2025
All rights reserved

UQÀM | Département des sciences
juridiques

Production Team

Copy editing	Tanina Drvar
Proofreading	Robbie McCaw
Typesetting	Nord Compo
Cover design	Benoit Denault
Cover Image	Pauline Stive

Centre for Environmental Law
and Global Sustainability

Centre du droit de l'environnement
et de la durabilité mondiale

 uOttawa

PUO-UOP gratefully acknowledges the funding support of the University of Ottawa, the Government of Canada, the Canada Council for the Arts, the Ontario Arts Council and the Government of Ontario.

Table of Contents

List of Figures.. xi
List of Tables xiii
List of Abbreviations.. xv

INTRODUCTION... 1
Thomas Burelli, Alexandre Lillo, Elie Klee,
Lynda Hubert Ta and Lauren Touchant

   1. 30 Years On, Time to Take Stock........................... 1
   2. A Canadian Perspective 6
   3. Recruiting Contributors to the Book 10
   4. Organization of the Book 13

CHAPTER 1
COP 101: Understanding What Climate COPs Are 15
Lynda Hubert Ta, Elie Klee, Thomas Burelli,
Alexandre Lillo, and Lauren Touchant

   1.1. Supreme Decision-Making Bodies to Help
        Implement International Climate Agreements
        and Achieve Their Objectives............................... 16
   1.2. Decisions of Varying Importance,
        Including Founding Agreements 19
   1.3. A Fair and Collaborative Organization,
        A Continuous Negotiation Process................... 21
   1.4. A Diversified and Growing Participation 24

    1.5. Efficiency and Relevance Called
          Into Question 31

CHAPTER 2

History of the COP from COP1 to COP30...................... 39
Jean Lemire

    2.1. Scientific Evidence... 39
    2.2. The Beginnings of Environmental Diplomacy.... 43
    2.3. Common but Differentiated Responsibilities.... 45
    2.4. Conferences of the Parties
          and the Kyoto Protocol....................................... 46
    2.5. The Paris Agreement.. 54
    2.6. The Road to Belém 73

CHAPTER 3

So COP, What Have You Done with Your Life? 91
Richard Kinley

    3.1. Let's Take a Step Back 92
    3.2. What Is the COP Really? 93
    3.3. A Thematic Chronology of the COPs............... 98
    3.4. Approaching COP30... 102
    3.5. How Could COPs Help Governments Be
          More Ambitious?... 104
    3.6. Conclusion.. 111

CHAPTER 4

A Fair, Funded, and Fossil-Free Future:
A Parliamentarian's Role in International
Climate Action... 113
Rosa Galvez

    4.1. My Introduction to COP 115
    4.2. Nomination to the Senate................................... 118
    4.3. COP as a Parliamentarian: A New Experience.... 121
    4.4. COP26: A Turning Point for Climate Finance.... 124

4.5. Attending COP27 and COP28............................ 127
4.6. Moving Forward: The Need for
        Greater Ambition 130
4.7. Conclusion.. 132

CHAPTER 5

COPs Are Important Gatherings:
They Keep Climate Change on the Map 135
Elizabeth May

CHAPTER 6

What It Takes to Represent Canada at Climate COPs:
A Federal Government Perspective 149
Catherine Stewart

6.1. Let Us Begin with Formal COP Negotiations.... 151
6.2. Moving Beyond Formal COP Negotiations..... 158
6.3. Then There Is the Programming Side
        of Things 159
6.4. A Gathering That Shapes Our Activities
        Throughout the Year... 160
6.5. There Is Also What We Take Away
        from COP, the Commitments We Make,
        and How We Apply Them at Home 162

CHAPTER 7

What Now? A Millennial's Take on a Decade of COPs... 165
Dominique Souris

7.1. The Beginning.. 167
7.2. Navigating the "Age of Bullshit"......................... 170
7.3. Building and Belonging in the UNFCCC......... 172
7.4. Building a Future Beyond.................................... 175

CHAPTER 8

Advancing Decarbonization and Decolonization:
Lessons from Indigenous Peoples Participation
in the United Nations Framework Convention
on Climate Change 179
Graeme Reed

    8.1. Opening Words 179
    8.2. Breathing Life into the FWG and
          the Platform 186
    8.3. Lessons from the Creation of FWG
          for the Future of the UNFCCC 197
    8.4. Conclusion 200

CHAPTER 9

ᓯᓚ ᐊᔾᔨᐅᔪᓐᓇᐃᖅᑐᖅ—Sila Ajjiujunnaiqtuq—
The Weather Has Changed: Inuit Perspectives and
Experiences from 30 Years of Climate Change COPs 201
Lisa Qiluqqi Koperqualuk, Sara Olsvig, Piita Irniq,
Alexina Kublu, Dalee Sambo Dorough,
Sheila Watt-Cloutier, Miyuki Qiajunnguaq Daorana,
Susie-Ann Kudluk, and Anne Simpson

    9.1. Inuit Voices and the Fight for Climate Justice 202
    9.2. Inuit Action Over the Last 30 Years 205
    9.3. Where We Are 209
    9.4. The Future of Inuit Advocacy at
          Climate Change COPs: Youth Perspectives 213
    9.5. Moving Forward 215

CHAPTER 10

Otipemisiwak and Climate Leadership:
The Métis Nation at UNFCCC 217
Dane de Souza and Kate Gillis

CHAPTER 11

The UNFCCC COP: An Imperfect—and Essential—
Civic Space for Climate Multilateralism 227
Caroline Brouillette

CHAPTER 12

Municipalities on the Front Line: Local Voices
in Global Climate Policy 239
Berry Vrbanovic and Lauren Touchant

CHAPTER 13

COPs Are Slow, But Cities Are on the Go 251
David Miller and Lauren Touchant

CHAPTER 14

From 2015 to COVID, We Were Still Dreaming 261
Patrick Rondeau

    14.1. A Crisis of Confidence 270
    14.2. The Road to Belém 276

CHAPTER 15

The UN Climate Conferences: A Driver for
"Impact Intrapreneurship" 279
Christophe Aura

    15.1. My Path to Intrapreneurship 280
    15.2. An Environment Favourable
        to Improbable Connections 283
    15.3. Developing a Sense of Shared Responsibility.... 289
    15.4. The COPs' Legacy on Impact
        Intrapreneurship: The Feeling of Being Part
        of Something Bigger Than Ourselves 293

CHAPTER 16

The Business of COP: How the Corporate Sector
Took on Climate... 295
John Stackhouse

    16.1. A Personal Journey Through COPs................ 298
    16.2. The Twin Journeys of COP and Commerce.... 301
    16.3. The Economy, Never Far from Sight............. 309

CHAPTER 17

Good COP, Bad COP: 20 Years of Climate Change
Negotiations... 313
Mark Purdon

    17.1. Kyoto Dreamin'..................................... 314
    17.2. Lessons Learned About COP........................... 317
    17.3. The Shift from Liberal Environmentalism
           to Developmental Environmentalism............. 322
    17.4. Towards Liberal Developmental
           Environmentalism?... 328
    17.5 Conclusion.. 330

List of Contributors 365

List of Figures

Figure 2.1.   Letter from the Brazilian presidency
of COP30   81

Figure 8.1.   A Short Overview of the Platform   187

Figure 17.1.   ODA for climate and non-climate
purposes, according to data from OECD
Development Assistance Committee
(constant 2022 USD)   327

List of Tables

Table 17.1. Characteristics of liberal environmentalism, developmental environmentalism and liberal developmental environmentalism 328

Table COP1 COP29 333

List of Abbreviations

AAU: Assigned Amount Unit
ACIA: Arctic Climate Impact Assessment
ADNOC: Abu Dhabi National Oil Company
AFN: Assembly of First Nations
AGBM: Ad Hoc Group on the Berlin Mandate
AOSIS: Alliance of Small Island States
AR4: Fourth Assessment Report (IPCC)
AWG-KP: Ad Hoc Working Group on Further Commitments
AWG-LCA: Ad Hoc Working Group on Long-Term Cooperative
    Action
BINGO: Business and Industry Non-Governmental Organizations
BOGA: Beyond Oil and Gas
BRIC: Brazil, Russia, India, and China
BRICS: Group formed by 11 countries: Brazil, Russia, India, China,
    South Africa, Saudi Arabia, Egypt, the United Arab Emirates,
    Ethiopia, Indonesia, and Iran.
CAN: Climate Action Network International
CAN-Rac: Climate Action Network Canada/Réseau action climat
CBAM: Carbon Border Adjustment Mechanism
CCS: Carbon Capture and Storage
CDM: Clean Development Mechanism
CEO: Corporate Europe Observatory
CERs: Certified Reduction Units
CGE: Consultative Group of Experts

CIEL: Centre for International Environmental Law
CMA: Conference of the Parties serving as the meeting of the Parties to the Paris Agreement
CMP: Conference of the Parties serving as the meeting of the Parties to the Kyoto Protocol
COP: Conference of the Parties
CRFTQMM: FTQ Montreal Metropolitan Regional Council
CTCN: Climate Technology Centre and Network
EGTT: Expert Group on Technology Transfer
EITs: Economies in Transition
ENGO: Environmental Non-Governmental Organizations
ERUs: Emission Reduction Units
ET: Emissions Trading
FAO: Food and Agriculture Organization
FCM: Federation of Canadian Municipalities
FSC: Forest Stewardship Council
FTQ: Fédération des travailleurs et des travailleuses du Québec
FWG: Facilitative Working Group
G7: Group of 7 (consisting of Canada, France, Germany, Italy, Japan, the United Kingdom and the United States)
G20: Group of 20 (consisting of Argentina, Australia, Brazil, Canada, China, France, Germany, India, Indonesia, Italy, Japan, Mexico, Russia, Saudi Arabia, South Africa, South Korea, Turkey, the United Kingdom, and the United States, along with the European Union and the African Union)
GCOS: Global Climate Observation System
GCF: Green Climate Fund
GEF: Global Environment Facility
GFANZ: Glasgow Financial Alliance for Net Zero
GGA: Global Goal on Adaptation
GHG: Greenhouse Gas
GPG: Good Practice Guidance
GW: Global Witness
HFCs: Hydrofluorocarbons
HWP: Harvested Wood Product
IAS: Invasive Alien Species
IC3: Faculty of Environment and the Waterloo Climate Institute
ICAO: International Civil Aviation Organization
ICC: Inuit Circumpolar Council

IEA: International Energy Agency

IIPFCC: International Indigenous Peoples Forum on Climate Change

IITC: International Indian Treaty Council

ILO: International Labour Organization

IMO: International Maritime Organization

INDC: Intended Nationally Determined Contributor

IPCC: Intergovernmental Panel on Climate Change

IPO: Indigenous Peoples organizations

IPU: International Parliamentary Union

IRP: Indigenous research paradigm

ITL: International Transaction Log

ITMO: Internationally Transferred Mitigation Outcomes

ITUC: International Trade Union Confederation

IWGIA: International Working Group on Indigenous Affairs

JI: Joint Implementation

JISC: Joint Implementation Supervisory Committee

LCERs: Long-term Certified Emission Reductions

LCIPP: Local Communities and Indigenous Peoples Platform

LDCs: Least Developed Countries

LDCF: Least Developed Countries Fund

LEG: Least Developed Countries Expert Group

LGMA: Local Governments and Municipal Authorities

LULUCF: Land Use, Land Use Change, and Forestry

MDB: Multilateral Development Bank

MDG: Millennium Development Goals

MP: Member of Parliament

MRV: Monitoring, Reporting, and Verification

NAMAs: Nationally Appropriate Mitigation Actions

NCQG: New Collective Quantified Goal

NDC: Nationally Determined Contributor

NGO: Non-Governmental Organization

NIO: National Indigenous Organization

ODA: Official Developmental Assistance

OECD: Organisation for Economic Co-operation and Development

PAICC: Paris Agreement Implementation and Compliance Committee

PAWP: Paris Agreement Work Program

PCCB: Paris Committee on Capacity Building

PFCs: Perfluorocarbons

PNCCS: Parliamentary Network on Climate Change and Sustainability

PPCA: Powering Past Coal Alliance

PSF: People's Social Forum

RINGO: Research and Independent Non-Governmental Organizations

SAR: Second Assessment Report (IPCCC)

SCCF: Special Climate Change Fund

SDG: Sustainable Development Goal

SBI: Subsidiary Body for Implementation

SST: Seychelles Support Team

TAR: Third Assessment Report (IPCC)

TCERs: Temporary Certified Reduction Units

TEC: Technology Executive Committee

TUNGO: Trade Union Non-Governmental Organization

UCLG: United Cities and Local Governments

UNCBD: United Nations Convention on Biological Diversity

UNCCD: United Nations Convention to Combat Desertification

UNCED: United Nations Conference on Environment and Development

UNCTAD: United Nations Conference on Trade and Development

UNDP: United Nations Development Programme

UNDRIP: United Nations Declaration on the Rights of Indigenous Peoples

UNEP: United Nations Environment Programme

UNFCCC: United Nations Framework Convention on Climate Change

USAID: United States Agency for International Development

WCI: Western Climate Institute

WEF: World Economic Forum

WGC: Women and Gender Constituency

WIM: Warsaw International Mechanism

WHO: World Health Organization

WSF: World Social Forum

WTO: World Trade Organization

WWF: World Wildlife Fund

YOUNGO: Youth Non-Governmental Organization

Introduction

*Thomas Burelli, Alexandre Lillo, Elie Klee,
Lynda Hubert Ta, and Lauren Touchant*

1. 30 Years On, Time to Take Stock

November 2025 marks a major milestone in the history of the fight against climate change. This is the moment when the international community will "celebrate" the 30th edition of the Conference of the Parties (COP) to the United Nations Framework Convention on Climate Change. This international agreement is also blowing out its 33rd candle.

Thirty times, state delegations from over 190 countries have met in an attempt to curb climate change. *Thirty times*, the decisions taken have proved insufficient in the face of the scale and complexity imposed by climate change. *Thirty times*, major compromises have been made—often to the benefit of Western economic models dependent on fossil fuels— by the international community.

Today, scientific forecasts are ever more alarmist, and the essential drastic societal changes that could trigger hope continue to be postponed. Let us be clear: We are slowly approaching, day after day, COP after COP, an apocalyptic scenario worthy of the—worst?—Hollywood movies. This reality is already a fact for certain regions of the world that are not fortunate enough to live sufficiently above sea level or to have the financial, technological, and material resources to protect themselves.

What does climate science tell us? Global temperatures have risen by 1.09°C since the pre-industrial era,[1] and human activities have unequivocally caused this increase.[2] Moreover, this warming is accelerating. In the first two decades of the twenty-first century, global surface temperature was 0.99°C higher than in the 1850–1900 period.[3] In 2024, the global temperature had risen for the first time to 1.54°C above pre-industrial levels.[4] The rise in global temperature is intrinsically linked to the increase in greenhouse gas (GHG) into the atmosphere. By trapping solar radiation in the atmosphere, these gases warm the Earth's surface and transform global climate cycles.[5] The problem? The more those gases accumulate, the greater the greenhouse effect. And they have accumulated greatly over recent decades due to human activities: Their concentration has increased by 54% since 1990 and by 12% between 2010 and 2023.[6] Year on year, the levels of GHGs

1. Intergovernmental Panel on Climate Change, "Summary for Policymakers," in *Climate Change 2023: Synthesis Report. Contribution of Working Groups I, II and III to the Sixth Assessment Report of the Intergovernmental Panel on Climate Change*, eds. Hoesung Lee and José Romero (IPCC, 2023), 4, https://www.ipcc.ch/report/ar6/syr/downloads/report/IPCC_AR6_SYR_SPM.pdf.

2. Intergovernmental Panel on Climate Change, "Summary for Policymakers," 4.

3. Intergovernmental Panel on Climate Change, "Summary for Policymakers," 4.

4. World Meteorological Organization, *State of the Climate 2024 - Update for COP29* (World Meteorological Association, 2024), ii–1, https://mural.maynoothuniversity.ie/19179/1/State-Climate-2024-Update-COP29_en.pdf.

5. It is important to differentiate between climate and weather. Weather is characterized by rapid, short-term, local variations in the lower layers of the atmosphere. Climate, on the other hand, is characterized by slow, long-term regional or global effects throughout the atmosphere. See: UNEP, *Climate Change 2001: The Scientific Basis* (Cambridge University Press, 2001), 87, https://www.ipcc.ch/site/assets/uploads/2018/03/WGI_TAR_full_report.pdf.

6. UNEP, *Climate Change 2001*, 87.

recorded in the atmosphere beat the previous year's record. The levels of carbon dioxide—the main GHG—recorded in 2019 have not been this high for at least two million years.[7] Yes, that is correct: *two million years.*

The rapid and unprecedented rise in global temperature is also putting considerable pressure on the international community and the commitments it has made in recent years. Indeed, the Paris Agreement sets a dual objective. The most ambitious one aims to limit the rise in global temperature to 1.5°C above pre-industrial levels by 2100—which would already have permanent effects and transform the daily lives of humanity. The least ambitious seeks to contain the rise in global temperature "well below 2°C above pre-industrial levels."[8] To achieve the most ambitious goal set by the Paris Agreement, the Intergovernmental Panel on Climate Change (IPCC) indicates that GHG emissions will need to be reduced by 43% below 2019 levels by 2030, and that carbon neutrality must be achieved by 2050.[9] As per the least ambitious target, a 21% reduction in GHG emissions must be achieved by 2030 compared with 2019 levels.[10] Even this most unambitious target remains a gigantic undertaking. Global net GHG emissions have increased by 12% since 2010,[11] and the most recent projections at the time of writing predict an increase of 3.1°C above pre-industrial levels by the end of the century.[12]

---

7. UNEP, *Climate Change 2001*, 87.
8. *Paris Agreement*, 12 December 2015, 3156 UNTS 79, art 2, 1. a) (entered into force 4 November 2016).
9. Intergovernmental Panel on Climate Change, "Summary for Policymakers," 21.
10. Intergovernmental Panel on Climate Change, "Summary for Policymakers," 21.
11. Intergovernmental Panel on Climate Change, "Summary for Policymakers," 21.
12. UNEP, *Emissions Gap Report 2024: No More Hot Air … Please! With a Massive Gap between Rhetoric and Reality, Countries Draft New Climate Commitments* (UNEP, 2024), 33.

Canada is no exception to this climate upheaval. Although trends vary across the country, the average annual temperature has risen almost twice as fast as the global average.[13] Admittedly, GHG emissions across Canada, all provinces and territories combined, fell by 8.5% from 2005.[14] Nevertheless, we are still a long way from the Canadian government's 2021 commitment to reduce GHG emissions by 40%-45% below 2005 levels by 2030.[15] Considerable efforts will have to be made quickly by Canada, even though its economy still relies heavily on fossil fuels, the main industry behind GHG emissions.[16]

While this context suggests the worst, international law and policy on climate change have changed dramatically over the last 30 years. In addition to consolidating around international conventions of near-universal scope, international law and policy have also become more complex, with the development of a multitude of strategies, tools, and instruments. Climate COPs embody this transformation in that they have become essential forums of the international legal order, and

---

13. Environment and Climate Change Canada, *Canadian Environmental Sustainability Indicators: Temperature Change in Canada* (Environment and Climate Change Canada, 2024), 6, https://www.canada.ca/content/dam/eccc/documents/pdf/cesindicators/temperature-change/2024/temperature-change-en.pdf.

14. Environment and Climate Change Canada, *National Inventory Report 1990-2023: Greenhouse Gas Sources and Sinks in Canada* (Environment and Climate Change Canada, 2025), https://publications.gc.ca/collections/collection_2025/eccc/En81-4-2023-1-eng.pdf.

15. Government of Canada, *Canada's 2021 Nationally Determined Contribution under the Paris Agreement* (Government of Canada, 2021), 1, https://unfccc.int/sites/default/files/NDC/2022-06/Canada%27s%20Enhanced%20NDC%20Submission1_FINAL%20EN.pdf.

16. In 2022, the energy sector accounted for 82% of Canada's total GHG emissions (Environment and Climate Change Canada, *National Inventory Report,* 38), including oil and gas, which accounted for 31% of total GHG emissions (Environment and Climate Change Canada, *Canadian Environmental Sustainability Indicators,* 9).

spaces where relations of political power and influence are predominant.

And yet, though we frequently hear about COPs in the media, few people know exactly what they are, or what goes on inside them. These COPs are eminently complex. They are made up of a multitude of bodies, they lead to numerous decisions, and they mobilize a plethora of acronyms. It is difficult to navigate the sheer volume of information churned out and disseminated by the COPs, just as it is arduous to grasp all the nuances of what goes on there, even for specialists. The nebulous nature of the COPs is the main reason why we wanted to produce this book and compile the views of stakeholders who have taken part in numerous climate COPs. This forum, so important for humanity, needs to be better known and more accessible to the general public.

Despite the proliferation and progress of international environmental law over the last 30 years, we have to be realistic. If this field continues to develop and the COPs continue to take place every year, it is still mainly because they have not produced the expected effects. International law on climate change and its COPs are often described as insufficiently binding. We are therefore faced with a paradoxical situation in which international climate change law is an abundant, complex, and evolving corpus, of undoubted interest from an intellectual point of view, but whose effects remain largely disappointing in the face of the objectives for which it was originally conceived. This is the second reason behind this book. The 30th COP on climate is an important milestone, not only for taking stock of the international political system, but also for asking why these COPs are still taking place, and for how much longer they might continue to do so.

In short, this volume seeks to make climate COPs more accessible by sharing the visions and experiences of individuals who have lived through the realities of this forum. It also attempts to facilitate understanding of the dynamics surrounding the COPs and, to some extent, their nature as forums for compromise. Additionally, these chapters here explore the future of the COPs, of what they could or should be. It is therefore aimed at anyone with an interest in international climate law and policy, anyone with a curiosity to better understand the climate COPs, anyone who wants to supplement their technical knowledge, anyone who wishes to better navigate the nuances and complexities of this forum, and finally anyone who, because of their eco-anxiety, would like to better equip themselves to decipher the reality in which we are evolving.

2. A Canadian Perspective

On the occasion of COP30 in Belém (Brazil), this book offers readers a Canadian perspective on the climate COPs that have taken place since 1995. But why take an interest in the Canadian perspective on climate COPs? First, because we are a team based in Canada, so what could be more logical than to start by taking an interest in what is happening here? But that is not enough, and there are many additional reasons.

Second, Canada is not the low contributor of GHG emissions that some statistics might lead you to believe. Indeed, given its current annual contribution—1.40% of global emissions in 2021[17]—Canada may appear to be a minor player compared to other countries, such as China, which will account for 27.9% of global emissions in 2021.[18] However, this figure does not give us a picture of Canada's qualitative and historical contribution to global GHG emissions.

On the one hand, Canada is one of the world's biggest contributors, with 17.7 tonnes of carbon dioxide equivalent per capita,[19] well below the per capita emissions of the

---

17. "Global Greenhouse Gas Emissions," Environment and Climate Change Canada, last modified March 6, 2025, https://www.canada.ca/en/environment-climate-change/services/environmental-indicators/global-greenhouse-gas-emissions.html.
18. "Global Greenhouse Gas Emissions."
19. "Global Greenhouse Gas Emissions."

Chinese.[20] On the other hand, Canada is among the top 10 historical emitters of GHGs, with a share of 2.6%.[21] In this sense, Canada has a particular and important responsibility when it comes to GHG emissions that contribute to climate change.[22] Given these figures, it is easy to understand why it is important to consider the role and actions of this country and its representatives in the negotiations.

What is more, Canada is a member of the Group of 7 (G7), which brings together the seven largest industrialized economies, as well as a member of the Group of 20 (G20), which brings together 19 of the world's most developed economies, the European Union, and the African Union. This status and the resources at Canada's disposal give it a responsibility in the fight against climate change.

Canada has far more resources (human, financial, and technological) than a country such as Tuvalu, for example, to contribute to the fight against climate change.[23] Some organizations are, therefore, insisting on what Canada's fair share[24]

---

20. Less than 10 tonnes per year per inhabitant: "Global Greenhouse Gas Emissions."
21. Simon Evans, "Analysis: Which Countries Are Historically Responsible for Climate Change?" *Carbon Brief*, October 5, 2021, https://www.carbonbrief.org/analysis-which-countries-are-historically-responsible-for-climate-change/.
22. Jodi Stark and Theresa Beer, "With Only 2% of Global Emissions, Why Does Canada's Climate Action Matter?" David Suzuki Foundation, July 24, 2024, https://davidsuzuki.org/expert-article/with-only-2-per-cent-of-global-emissions-why-does-canadas-climate-action-matter/.
23. In 2020, Tuvalu had a population of less than 12,000 and its GDP in 2019 was less than $50 million.
24. This fair share takes into account Canada's current and historical responsibility, as well as its capacity to act in the fight against climate change. Climate Action Network Canada, *Canada's Fair Share towards Limiting Global Warming to 1.5°C* (Climate Action Network Canada, 2019), https://climateactionnetwork.ca/wp-content/uploads/2019/12/Canada-Fair-Share-Infographic.pdf.

should be when it comes to reducing GHG emissions on a global scale.[25]

Beyond its contribution to GHG emissions and its capacity for action, Canada deserves our attention at the climate COPs because of its active participation in these meetings. Although this participation has not always been a smooth ride,[26] Canada has historically been a major participant in the climate COPs. Since the first COP, 2,407 individuals have taken part in these international meetings as part of the Canadian delegation;[27] 25 in 1995 at the first COP, 51 at the COP that saw the birth of the Kyoto Protocol (COP3), 287 at COP21 in Paris, and finally 636 members, the largest Canadian delegation, at COP28 in 2023 in Dubai. Since the Paris Agreement, an average of 226 members of the Canadian

---

25. About Canada's fair share: Climate Action Network Canada, *Canada's Fair Share*. Quebec's fair share: Climate Action Network Canada, *La juste part du Québec pour limiter le réchauffement mondial à 1.5°C* (Climate Action Network Canada, 2019), https://climate actionnetwork.ca/wp-content/uploads/2021/10/21-23-FairShare-Infographic_Quebec_OCT2021.pdf. See also: Karine Péloffy and Nick Zrinyi, "Canada's Fair Share of Emissions Reductions under the Paris Agreement," April 2021, https://rosagalvez.ca/en/initiatives/climate-accountability/canada-s-fair-share-of-emissions-reductions-under-the-paris-agreement/.

26. Canada was the first G7 country to ratify the 1992 Convention, was one of the first signatories of the Kyoto Protocol in 1998, eventually withdrew from it, and is now a strong supporter of the Paris Agreement. Silvia Maciunas and Géraud de Lassus Saint-Geniès, "The Evolution of Canada's International and Domestic Climate Policy from Divergence to Consistency?" *Canada in International Law at 150 and Beyond*, Paper No. 21, 2018, https://www.cigionline.org/static/documents/documents/Reflections%20Series%20Paper%20no.21%20Maciunas.pdf.

27. Based on lists of COP participants provided by the UNFCCC Secretariat we have created our own database. These are single participants. Including multiple participations by certain individuals, Canada sent 2,981 people to the COPs as part of its official delegation since the first COP.

delegation have taken part in the COPs each year. The size of the delegation has increased over the years,[28] as has the total number of participants in the climate COPs.[29] This tends to demonstrate the growing importance of these forums on the international stage.

In addition to the size of its delegation, Canada has often been seen as an important player in terms of its statements and its commitment to international negotiations. Despite some setbacks, such as its withdrawal from the Kyoto Protocol, Canada has generally played a leading role in international climate negotiations and, since the Paris Agreement in 2015, has shown a willingness to make progress and contribute to the success of important international agreements. In 2016, the Canadian Prime Minister Justin Trudeau declared at the signing of the Paris Agreement: "Today, with my signature, I give you our word that Canada's efforts will not cease."[30] Since then, Canada has regularly reaffirmed its commitment to the Paris Agreement.[31]

---

28. However, the increase in size is not linear. For example, after 207 participants at COP15, the Canadian delegation did not exceed 100 members until COP21 in Paris. At COP29 in Baku in 2024, the delegation was one of the smallest since 2013, with 62 members.

29. Robert McSweeney, "Analysis: How Delegations at COP Climate Summits Have Changed Over Time," *Carbon Brief*, October 27, 2021, https://www.carbonbrief.org/analysis-how-delegations-at-cop-climate-summits-have-changed-over-time/.

30. Catherine Cullen, "Justin Trudeau Signs Paris Climate Treaty at UN, Vows to Harness Renewable Energy," *CBC News*, April 22, 2016, https://www.cbc.ca/news/politics/paris-agreement-trudeau-sign-1.3547822.

31. However, the translation of this desire into concrete results at national level is questionable; see the Environment Commissioner's reports. It should be noted, however, that the federal government, which is responsible for negotiating international agreements, is not the only one competent to combat climate change at the national level. In Canada, jurisdiction over the environment is shared between the federal government, the provinces, the territories, and the municipalities.

3. Recruiting Contributors to the Book

Although we know that several hundred Canadians had taken part in the climate COPs since 1992, how could we identify potential contributors to this book? To answer this question, we employed several strategies.

First, we delved into the past of the COPs and built a database of participants in the Canadian delegation since the first climate COP. To do this, we accessed the lists of COP participants published by the Secretariat of the United Nations Framework Convention on Climate Change at each COP. These lists specify the composition of the state delegations taking part in the COPs. These are the participants who are part of the official state delegation and who have been given "party" accreditation. They do not include observers of a given nationality, who may also take part in the COPs in large numbers, representing organizations accredited by the Secretariat (often more than the members of a state delegation). Given the scale of the task, we have only constructed a database of the members of the official Canadian delegation from COP1 to COP 29. The information provided by the Secretariat of the Framework Convention also gave us access to the status and function of the participants in the Canadian delegation. We were able to observe the presence of a wide variety of players within the Canadian delegation: negotiators, industry representatives, representatives of the many branches of civil society (youth, environment, etc.), members of provincial governments, First Nations representatives, as well as IT support staff and drivers. Finally, this strategy enabled us to identify 2,407 people who had participated in one or more COPs from COP1 in 1995 to COP29 in 2024. We then used this database to contact potential contributors.

We also activated our networks and relied on word-of-mouth to identify relevant contributors for this book. We believed that some individual stories would be of great interest to the public and decided to contact them. We also asked the people we contacted to recommend other potential contributors.

Finally, we selected the contributors using two main criteria: the participation of contributors in the COPs and their status. Firstly, their involvement in the COPs. We gave preference to people who had taken part in several COPs and had a wealth of experience. However, we did not automatically disqualify people with more limited experience of COPs.

One of the challenges has been the age of the COP process and the fact that some contributors and potential contributors were no longer active or able to be contacted since their participation in the COPs. This is particularly true of those who took part in the first COPs from 1995 onwards. Thus, while some contributors took part in COPs some time ago, others took part in COPs in the 1990s, while others attended in the 2020s. This contributes to the diversity of perspectives offered to readers in this book.

We then invited contributors to participate according to their stakeholder category when they attended the COPs. In this way, we tried to give readers a picture of the diversity of Canadian players taking part in the climate COPs. To this end, we contacted potential contributors in the following sectors:

- United Nations Framework Convention on Climate Change (UNFCCC) bodies
- Canadian federal government/federal level
- Provincial and territorial governments
- Municipal governments/local authorities
- Indigenous Peoples
- Environmental NGOs
- Youth
- Trade unions/workers and their unions
- Trade and industry/private sector (natural resources and finance)
- Scientific and technical community scientists (social and climate sciences)

We identified these categories on the basis of the list of people who took part in the Canadian delegation and the types of non-state actors provided for in Agenda 21.[32] All these elements led to the following list of contributors (presented below in alphabetical order):

- Christophe Aura, President at COPTICOM
- Caroline Brouillette, Executive Director of Climate Action Network Canada
- Miyuki Qiajunnguaq Daorana, MA Candidate and Inuit Youth Activist
- Dane de Souza, Senior Advisor on Emergency Management at Métis Nation Council
- Dalee Sambo Dorough, Iñupiaq Advocate and Lawyer
- The Honourable Rosa Galvez, Senator
- Kate Gillis, Senior Policy Advisor, International Relations at Métis National Council
- Piita Irniq, Elder and Knowledge Holder
- Richard Kinley, Deputy Executive Secretary at the UN Climate Change Secretariat from 2006 to 2017
- Lisa Qiluqqi Koperqualuk, Vice-Chair, Inuit Circumpolar Council
- Alexina Kublu, Elder and Knowledge Holder
- Susie-Ann Kudluk, National Inuit Youth Council President
- Jean Lemire, Quebec's Climate Change Envoy
- Elizabeth May, Member of Parliament
- David Miller, Managing Director, C40 Centre for City Climate Policy and Economy
- Erin Myers, Director of Environment and Climate Change at Métis National Council
- Sara Olsvig, Chair, Inuit Circumpolar Council
- Mark Purdon, Professor, Department of Strategy, Social and Environmental Responsibility, Holder, Chair in

---

32. United Nations Sustainable Development, "Agenda 21," United Nations Conference on Environment & Development, Rio de Janeiro, Brazil, June 3 to 14, 1992, https://sustainabledevelopment.un.org/content/documents/Agenda21.pdf.

Decarbonization, École des sciences de la gestion, University of Quebec at Montréal

- Graeme Reed, Strategic Advisor at the Assembly of First Nations
- Patrick Rondeau, Environmental and Just Transition Department Director, Fédération des travailleurs et des travailleuses du Québec (FTQ)
- Anne Simpson, Inuit Circumpolar Council Climate Change Advisor
- Dominique Souris, Social Entrepreneur and Impact Strategist
- John Stackhouse, Senior Vice President, Office of the CEO, Royal Bank of Canada
- Catherine Stewart, Canada's Former Ambassador for Climate Change
- Berry Vrbanovic, Mayor of Kitchener
- Sheila Watt-Cloutier, Inuk Leader, Kuujjuaq, Nunavik (Northern Quebec)

4. Organization of the Book

As part of the approach to consider how to organize the contributions we did not wish to rank the contributions in order of importance, which would imply a subjective classification. We, therefore, decided to arrange the chapters according to the most classic categories of international environmental law. The first two sections comprise contributions from representatives of formal actors, namely international organizations and member states. The following sections include chapters from representatives of non-state actors, organized according to Agenda 21 classification and order: Youth, First Nations, Métis, and Inuit, Non-Governmental Organizations, Local Authorities, Workers and Unions, Business and Industry, and Scientific Community.[33] Within those Agenda 21 categories, the chapters have been organized by alphabetical order. However, we want to highlight the fact that the contributions are not part of a linear overall narrative. On the contrary,

---

33. United Nations Sustainable Development, "Agenda 21."

they represent a diversity of perspectives, which, we hope, will enable readers to grasp the diversity of actors who converge and meet at the COPs, as well as their respective roles and contributions.

Climate COPs are complex events, and though we hear about them regularly in the media every year around November or December, they can be mysterious to the public (and sometimes even to students of international environmental law). It is hard not to get lost between the COPs, CMAs, CMPs and other acronyms, the dozens of sometimes lengthy decisions that are adopted at each COP, not to mention subjects as technical as the Warsaw International Mechanism for Loss and Damage associated with Climate Change Impacts or the Poznan strategic program on technology transfer. That is why we have decided to offer readers who are less familiar with the COPs a "COP 101" chapter to share some essential keys to understanding them.

We hope that this book will help you to better understand the COPs, to appreciate their gains as well as their limits. Perhaps it will encourage you to take part yourself and try to make a difference in the fight against climate change at the international level, as the contributors to this book have done by sharing their experiences. Above all, we hope that in 30 years' time we will not have to produce another review of 60 years of climate COPs.

CHAPTER 1

COP 101: Understanding What Climate COPs Are

Lynda Hubert Ta, Elie Klee, Thomas Burelli,
Alexandre Lillo, and Lauren Touchant

One might ask: Since treaties such as the United Nations Framework Convention on Climate Change (UNFCCC),[1] the Kyoto Protocol,[2] and the Paris Agreement already exist, what is the purpose of the Conferences of the Parties?[3] What is their role? Where and by whom are they organized? Who attends, and for what purpose? How do they function in general? In order to answer these questions, we will examine their function in the international climate regime, their organization, participants, and functioning.

---

1. *United Nations Framework Convention on Climate Change*, 9 May 1992, 1771 UNTS 107, https://unfccc.int/resource/docs/convkp/conveng.pdf.

2. *Kyoto Protocol to the United Nations Framework Convention on Climate Change*, 11 December 1997, 2003, UNTS 162, 37 ILM 22, https://unfccc.int/resource/docs/convkp/kpeng.pdf.

3. *Paris Agreement*, 12 December 2015, 3156 UNTS 79, (entered into force 4 November 2016), https://unfccc.int/sites/default/files/english_paris_agreement.pdf.

1.1. Supreme Decision-Making Bodies to Help Implement International Climate Agreements and Achieve Their Objectives

First, as De Lassus St-Geniès points out, "a United Nations climate conference is […] not, strictly speaking, just a 'COP.' It is in fact the meeting of three Conferences of the Parties" (our translation).[4] In fact, the UNFCCC, the Kyoto Protocol, and the Paris Agreement have their own Conference of the Parties, which is a conference that periodically brings together the states and international organizations that have ratified them (the "parties"). Each of these agreements establishes an institutional framework for international cooperation, enabling negotiations to continue. Far from being the end of the story, their adoption is just the beginning.

For efficiency, their respective Conferences of the Parties are held at the same time and place every year. To differentiate them, the annual meeting of the parties to the UNFCCC is called the "Conference of the Parties" (COP), that of the parties to the Kyoto Protocol is called the "Conference of the Parties serving as the meeting of the Parties to the Kyoto Protocol" (CMP), while that of the parties to the Paris Agreement is called the "Conference of the Parties serving as the meeting of the Parties to the Paris Agreement" (CMA). The first CMP (CMP1) was held in Montréal in 2005, at the same time as COP11. The first CMA (CMA1) was held in Marrakech in 2016 at the same time as COP22 and CMP12.

Although the Kyoto Protocol is no longer relevant since the adoption of the Paris Agreement, it "technically remains in force."[5] The CMPs therefore continue to take place, for example, to continue compiling and reviewing reports on the

---

4.   Géraud de Lassus St-Geniès, "Changements climatiques : à quoi servent les COP ?" *Le Climatoscope* (February 2020): 61, https://climatoscope.ca/wp-content/uploads/2020/02/DeLassus_Climatoscope.pdf.

5.   Daniel Bodansky, "Introductory Note," *Audiovisual Library of International Law*, July 2021, https://legal.un.org/avl/ha/pa/pa.html.

implementation of the Protocol by the various parties, and to discuss issues relating to the mechanisms it created, such as international emissions trading and the Clean Development Mechanism.[6] According to Collard, the carbon market established by the Protocol remains "essential" to the international climate regime, even if it needs to be reformed so as to be better adapted to the terms of the Paris Agreement, which, in effect, replaced the Protocol as the main legal instrument of that regime.[7] The international emissions trading system was included in Article 6 of the Paris Agreement, but the rules for its implementation were not specified until COP29 in Baku.[8] Interrelationships also remain between the Protocol and the two other agreements.

Secondly, it is important to note that the Conference of the Parties is first the supreme decision-making body of the international agreement that established it. As such, like the CMP of the Kyoto Protocol and the CMA of the Paris Agreement,[9] the COP of the UNFCCC "shall keep under regular review the implementation of the Convention […] and shall make […] the decisions necessary to promote [its] effective implementation."[10]

This annual meeting provides an opportunity to examine the application of these three international legal instruments

6. See for example *Guidance relating to the clean development mechanism*, UNFCCC, 19th Sess, 2024, UN Doc FCCC/KP/CMP/2024/L.3, Dec CMP.19 https://unfccc.int/sites/default/files/resource/CMP6 _CDM_guidance_Agenta_item_5_0.pdf.

7. Fabienne Collard, "Les COP sur les changements climatiques," *Courrier hebdomadaire du CRISP* 2486–2487, no. 1 (April 2021): 51, https:// doi.org/10.3917/cris.2486.0005.

8. UNFCCC, "COP29 UN Climate Conference Agrees to Triple Finance to Developing Countries, Protecting Lives and Livelihoods," announcement, November 24, 2024, https://unfccc.int/news/cop29-un-climate-conference-agrees-to-triple-finance-to-developing-countries-protecting-lives-and.

9. *Kyoto Protocol*, art 13 para 4.; *Paris Agreement*, art 16 para 4.

10. *Paris Agreement*, art 7 para 2.

by their parties, to assess the effectiveness of the measures taken, to evaluate the progress made and to make any necessary adjustments. For example, each party to the Paris Agreement has undertaken to submit a nationally determined contribution (NDC) forecasting its efforts to reduce national greenhouse gas (GHG) emissions, and to update it every five years to progressively raise its ambitions.[11] Every five years, the Conference of the Parties carries out a global review of progress towards achieving the objectives of this agreement, the first of which took place at COP28 in Dubai in 2023.[12] The outcome of this review guides the preparation of the next five-yearly NDCs to better align global climate action with the achievement of the agreement's objective.[13]

The Conference of the Parties is also an opportunity[14] to share information on the best measures, methods, and practices, enabling participants to compare and inspire each other. It is an opportunity to formulate recommendations, which can give impetus to new policies and partnerships. It is also an opportunity to mobilize the financial resources needed to achieve climate objectives. For example, at COP27 in Sharm el-Sheikh, Egypt, in 2022, it was decided to create a financial aid fund to enable those countries most vulnerable to the effects of climate change to cope with the losses and damages associated with these effects.[15] This fund was operationalized at the following COP in Dubai.[16]

---

11. *Paris Agreement*, art 4 para 2 and 9.
12. *Paris Agreement*, art 14 para 2.
13. *Paris Agreement*, art 14 para 3.
14. *United Nations Framework Convention on Climate Change*, art 7 para 2.
15. *Funding arrangements for responding to loss and damage associated with the adverse effects of climate change, including a focus on addressing loss and damage*, UNFCCC, 4th Sess, UN Doc FCCC/PA/CMA/2022/10/Add.1 (2022) Dec 2/CMA at pages 13–16, https://unfccc.int/sites/default/files/resource/cma2022_10_a01E.pdf.
16. *Operationalization of the new funding arrangements, including a fund, for responding to loss and damage referred to in paragraphs 2–3 of decisions 2/CP.27 and 2/CMA.4*, UNFCCC, 28th Sess, UN Doc FCCC/

Finally, the Conference of the Parties makes it possible to refine the provisions of the agreement that established it, in the light of experience gained during its application and of scientific advances.[17] This is how the decisions taken at COP3 and COP21 gave rise to the Kyoto Protocol and the Paris Agreement, two agreements that have enabled the UNFCCC to continue to develop and be implemented. The negotiations leading to the adoption of the Protocol, for example, were informed by the evolution of scientific knowledge contained in the Second Assessment Report of the Intergovernmental Panel on Climate Change (IPCC) in 1995, justifying its necessity and on which it was based.[18]

1.2. Decisions of Varying Importance, Including Founding Agreements

The elements negotiated and agreed by states at a Conference of the Parties are recorded in "decisions," adopted at the end of the conference. These decisions are numerous. By way of illustration, in Dubai in 2023, 19 decisions relating to COP28 were adopted, along with seven decisions concerning the CMP and 21 concerning the CMA. By the end of COP29 in Baku in 2024, 972 decisions had been adopted at all the COPs, CMPs, and CMAs combined.[19]

These decisions cover a wide range of issues and are of varying degrees of importance. Some are purely logistical, relating to the venue and date of the next COP,[20] to administrative

---

CP/2023/11/Add.1 (2023) Dec 1/CP.28 at pages 2–18, https://unfccc. int/sites/default/files/resource/cp2023_11a01E.pdf.

17.   *United Nations Framework Convention on Climate Change*, art 7 para 2(a).

18.   Fabienne Collard, "Les COP sur les changements climatiques," 35–36.

19.   UNFCCC, "Decisions," COP29, accessed March 28, 2025, https://unfccc.int/decisions.

20.   See for example: *Dates and venues of future sessions*, UNFCCC, UN Doc FCCC/CP/2017/11/Add.2 (2017) Dec 22/CP.23 at pages 26–27,

considerations,[21] to the creation of thematic work programs,[22] to the management of subsidiary bodies,[23] etc. Others are more substantive. It is through them that certain provisions of the UNFCCC and their application modalities can be refined, so as to further specify the obligations of the parties in implementing the Convention. This is how decisions taken at COP3 and COP21 led to the adoption of the Kyoto Protocol and the Paris Agreement, two major legally binding texts.

The legal scope of these decisions must be distinguished from the scope of the treaties adopted to implement them. For example, the scope of decision 1/CP.21, entitled "Adoption of the Paris Agreement,"[24] should not be confused with the scope of the Paris Agreement itself, which this decision establishes, and which appears in its annex.[25] Treaties, like the Paris Agreement, are formally binding. As for COP decisions, such as 1/CP.21, their scope is more ambiguous. Although not binding on the parties, that is not subject to sanctions in

---

    https://unfccc.int/sites/default/files/resource/docs/2017/cop23/eng/11a02.pdf.

21. See for example: *Administrative, financial and institutional matters*, UNFCCC, Un Doc FCCC/CP/2016/10/Add.2 (2016) Dec 23/CP.22 at pages 22–29, https://unfccc.int/sites/default/files/resource/docs/2016/cop22/eng/10a02.pdf.

22. See for example: *United Arab Emirates just transition work programme,* UNFCCC, UN Doc FCCC/PA/CMA/2023/16/Add.1 (2023) Dec 3/CMA.5 at pages 29–31, https://unfccc.int/sites/default/files/resource/cma2023_16a01E.pdf.

23. See for example: *Division of labour between the Subsidiary Body for Implementation and the Subsidiary Body for Scientific and Technological Advice*, UNFCCC, UN Doc FCCC/CP/1997/7/Add.1 (1997) Dec 13/CP.3 at pages 44–47, https://unfccc.int/sites/default/files/resource/docs/cop3/07a01.pdf.

24. *Adoption of the Paris Agreement*, UNFCCC, 21st Sess, UN Doc FCCC/CP/2015/10/Add.1 (2015) Dec 1/CP.21 at pages 2–20, https://unfccc.int/resource/docs/2015/cop21/eng/10a01.pdf.

25. *Adoption of the Paris Agreement*, UNFCCC, 21st Sess, UN Doc FCCC/CP/2015/10/Add.1 (2015) Dec 1/CP.21 at pages 21–36, https://unfccc.int/resource/docs/2015/cop21/eng/10a01.pdf.

the event of non-compliance, these decisions are nevertheless not without a certain normative value, notably practical and operative, producing effects on the behaviour of the parties.[26]

It is also worth noting that journalists and commentators generally focus almost exclusively on the first COP decision, which is often one of the most important—if not *the* most important—decision,[27] but not always.[28] On the other hand, the subsequent decisions, which may form the practical outgrowth of this first major decision by setting up its mechanisms and their financing, are just as interesting.

1.3. A Fair and Collaborative Organization, A Continuous Negotiation Process

Each year, the COP is hosted by a different state party, which then, in principle, holds the presidency. This host country is selected according to a process that strives to be equitable. Efforts are made to ensure geographical rotation between the five regional groups recognized by the United Nations (UN): Africa, Asia and the Pacific, Latin America and the Caribbean, Eastern Europe, Western Europe and others.[29] When it is a

---

26. Sandrine Maljean-Dubois, "L'accord de paris sur le climat, un renouvellement des formes d'engagement de l'État?" in *Quel(s) droit(s) pour les changements climatiques?*, ed. Marta Torre-Schaub, Christel Cournil, Sandra Lavorel, and Marianne Moliner-Dubost (Le Kremlin-Bicêtre: Mare et Martin, 2018), 5–6, https://shs.hal.science/halshs-02109672/document.

27. See for example: *Decision 1/CP.1 on the Berlin Mandate, Decision 1/CP.3 on the Adoption of the Kyoto Protocol, Decision 1/CP.21 on the Adoption of the Paris Agreement, Decision 1/CP.28 on the Operationalization of the new funding arrangements, including a fund, for responding to loss and damage referred to in paragraphs 2–3 of decisions 2/CP.27 and 2/CMA.4.*

28. An interesting example is COP29, where no framework decision was adopted.

29. UNFCCC, *Guide de la Convention-cadre des Nations unies sur les changements climatiques* (UNFCCC, 2008), 32, https://unfccc.int/resource/docs/publications/handbook_fr.pdf.

regional group's turn to host the COP in the coming year, the countries that make up or are linked to that group can submit their candidacy. They must make a concerted choice from among the proposed candidates. This choice is notified to the UNFCCC Secretariat, which is responsible for assessing the hosting capacity of the selected candidate. The successful candidate must have the logistical, technical, material, and financial capacity to host an international event of this scale, bringing together tens of thousands of participants. The Secretariat reports on these elements to the current year's COP Bureau. This choice must then be confirmed by the Subsidiary Body for Implementation (SBI), before it can be officially accepted by the COP.[30]

In practice, however, the hosting of the COP has not always been evenly distributed between the different regional groups. It was not until 2008 that a COP (COP14) was held for the first time in Eastern Europe, in Poznan, Poland. Moreover, to date, 12 of the 29 COPs have been organized in the "Western Europe and Others" region.[31] This could be partly explained by the fact that if no state party offers to host the COP, or if a state is offered but does not have sufficient capacity to take on such a large-scale event, then it is held in Bonn, Germany, at the headquarters of the UNFCCC Secretariat,[32] which has been the case on three occasions to date. For example, for logistical reasons, the presidency of

30. UNFCCC, *How to COP: A Handbook for Hosting United Nations Climate Change Conferences* (UNFCCC, 2023), 2–3, https://unfccc.int/sites/default/files/resource/How-to-COP.pdf.

31. In addition to continental Western Europe, this regional group includes Australia, Canada, Iceland, Israel, New Zealand, Turkey, and the United States.

32. UNFCCC, *United Nations Framework Convention on Climate Change Handbook* (UNFCCC, 2006), 28, https://unfccc.int/resource/docs/publications/handbook.pdf; *Organizational Matters: Adoption of the Rules of Procedure*, Rule 3: 3, UNFCCC, Un Doc FCCC/CP/1996/2, https://unfccc.int/sites/default/files/resource/02_0.pdf.

COP23 in 2017 was held by Fiji, while the conference took place in Bonn.

The UNFCCC Secretariat and the host country work hand in hand to organize the COP. From the outset, they sign a legally binding Host Country Agreement, which sets out the framework for the organization of the COP by the Secretariat in the host country, as well as their respective roles and obligations in its preparation, organization, and running.[33] For example, the Secretariat is responsible for the registration and accreditation of COP participants,[34] while the host country is responsible for providing the premises, equipment, facilities, and services (visas, accommodation, catering, transport, translation, offices and pavilions, support staff, information and communication technologies, etc.) required for its smooth running.[35] Another example: The host country takes care of protocol outside the conference site, while the Secretariat's protocol service takes care of protocol inside the site, such as coordinating the arrival of dignitaries, the opening ceremony and the high-level segment attended by heads of state and government, the UN secretary-general, ministers, etc.[36] The Secretariat and the host country also collaborate on the official programming of the conference, including social and cultural events and activities.[37] They also collaborate on securing the site. An agreement is signed to this effect, dividing up their security responsibilities. Generally speaking, while the host country ensures security outside the COP site, UN security personnel ensure security within the site's boundaries, and the host country is obliged to assist them.[38]

During the COP, the Secretariat, in conjunction with the host country, supports the negotiations by facilitating their

33. UNFCCC, *How to COP*, 8.
34. UNFCCC, *How to COP*, 14.
35. UNFCCC, *How to COP*, 16–28.
36. UNFCCC, *How to COP*, 10.
37. UNFCCC, *How to COP*, 12.
38. UNFCCC, *How to COP*, 29–30.

smooth running and ensuring that all the necessary support is in place. Between COPs, the Secretariat organizes and supports intersessional negotiations or preparatory meetings throughout the year, aimed at working on the major orientations to be presented at the COP. This is the case for each pre-COP, organized several weeks before a COP and bringing together the environment and energy ministers of some 50 countries, as well as representatives of each negotiating group. In this context, as in the case of the COPs, it is responsible for facilitating negotiations between countries. Throughout the year, it also organizes numerous other smaller meetings and workshops. The Secretariat, which employs some 450 people from over 100 countries,[39] plays a less publicized but no less essential role.

1.4. A Diversified and Growing Participation

1.4.1. *A Growing Number of Participants*

Every year, the COP climate brings together tens of thousands of people over a two-week period. It has not always attracted large crowds, but as a sign of the growing importance of this conference, the number of participants has gradually increased over time. This inflation has become particularly marked in recent years (around 10,000 participants at COP3 in Kyoto in 1997,[40] just under 40,000 at COP26 in Glasgow in 2021,[41] 49,000 at COP27 in Sharm el-Sheikh in 2022,[42] over

39. UNFCCC, "About the Secretariat: What Is the Purpose of the Secretariat," accessed March 28, 2025, https://unfccc.int/about-us/about-the-secretariat.

40. *List of Participants*, UNFCCC, 3rd Sess, UN Doc FCCC/CP/1997/INF.5 (1997), https://unfccc.int/sites/default/files/resource/docs/cop3/inf05.pdf.

41. *List of* Participants, UNFCCC, 26th Sess, UN Doc FCCC/CP/2021/INF.3 (2021), https://unfccc.int/sites/default/files/resource/cp2021_inf03p01_adv.pdf.

42. *List of Participants*, UNFCCC, 27th Sess, UN Doc FCCC/CP/2022/INF.3 (2022), https://unfccc.int/sites/default/files/resource/cp2022_inf03_part1.pdf.

80,000 at COP28 in Dubai,[43] and 55,000 at COP29 in Baku[44]).
Attendance peaked at the COPs that adopted the founding agreements: COP3 (around 10,000 people) and COP21
(around 30,000 people).[45]

1.4.2. State Delegations

As the COP is the annual meeting of the parties to the
UNFCCC, its first participants are the official delegations
of each state party. At COP29, held in Baku, Azerbaijan,
in November 2024, the Canadian delegation comprised
63 people for the main delegation and 204 people for the
overflow list.[46] The minister of the environment and climate
change, Steven Guilbeault, chaired the Canadian delegation
at COP29, with the support of Canada's ambassador for climate change, Catherine Stewart. Canada's chief negotiator
for climate change, Jeanne-Marie Huddleston, led Canada's
participation in the negotiations.

National delegations generally have a two-headed leadership: On the one hand, a head of delegation, whose role is
more political, who leads high-level bilateral or multilateral
meetings, and on the other, a chief negotiator who is familiar
with the complex mechanism of negotiations, coordinates

---

43. *List of Participants*, UNFCCC, 28th Sess, FCCC/CP/2023/INF.3 (2023),
    https://unfccc.int/sites/default/files/resource/cp2023_inf03.pdf.

44. UNFCCC, "COP 29 : la Conférence des Nations Unies sur le climat
    convient de tripler le financement aux pays en développement pour
    protéger les vies et les moyens de subsistance," accessed November 24,
    2024, https://unfccc.int/fr/news/cop-29-la-conference-des-nations-
    unies-sur-le-climat-convient-de-tripler-le-financement-aux-pays-en.

45. *List of Participants*, UNFCCC, 21st Sess, Un Doc FCCC/CP/2015/
    INF.3 (2015), https://unfccc.int/sites/default/files/resource/docs/2015/
    cop21/eng/inf03p01.pdf.

46. Government of Canada, "UN Conference on Climate Change: COP29
    in Azerbaijan," accessed December 16, 2024, https://www.canada.ca/en
    /services/environment/weather/climatechange/canada-international-
    action/un-climate-change-conference/cop29-summit.html.

the work of the negotiators, supports and accompanies them, receives daily reports from them, takes decisions in real time during negotiations, and reports to the head of delegation.[47]

In the case of the Canadian delegation, the minister of the environment and climate change is traditionally the head of the delegation, and the chief negotiator for the climate change file is drawn from the senior federal public service, usually Environment and Climate Change Canada. For example, Jeanne-Marie Huddleston, chief negotiator for Canada's delegation to COP29 in Baku, is director general of multilateral affairs and climate change at Environment and Climate Change Canada.[48]

1.4.3. *Canada at the COPs*

Canada's delegation to COP29 reflected the country's diversity. Many of its members were civil servants from various federal departments, principally Environment and Climate Change Canada and Global Affairs Canada, but also Innovation, Science and Economic Development Canada, Natural Resources Canada, Transport Canada, Health Canada, Agriculture and Agri-Food Canada, and Fisheries and Oceans Canada. These officials served as negotiators in various working groups (Agriculture, Loss and Damage, Adaptation, Mitigation, Gender and Capacity Building, Local Communities and Indigenous Peoples Platform, Transparency, Climate Finance) or as policy analysts, legal advisors, or coordinators.

47. Andrès Mogro, *Guide pour les négociateurs sur le changement climatique* (Ministerio des Ambiante, Agua y Transición Ecológica, 2021), 25, https://impulsouth.org/wp-content/uploads/2023/07/Guide-Pour-LesNegociateursSurLeChangementClimatique-fr.pdf.

48. Government of Canada, "Person Information: Jeanne-Marie Huddleston - Director General, MAAC and Chief Climate Negotiator," accessed October 10, 2020, https://geds-sage.gc.ca/fr/

The Canadian delegation also included representatives from the House of Commons and the Senate, provincial and territorial governments (including Quebec and Alberta), civil society organizations (such as Équiterre and Réseau environnement), universities (such as the University of Saskatchewan and the University of Calgary), companies (such as Rio Tinto Aluminium and ArcelorMittal), including a good number of oil companies (Pathways Alliance, a consortium of Canada's six largest oil sands developers, Imperial Oil, a Canadian subsidiary of ExxonMobil, etc.) and mining companies (Vale SA, ArcelorMittal Mining Canada), young people (including a leader of the Youth Climate Club), as well as Indigenous representatives (including members of the Assembly of First Nations, the Inuit Circumpolar Council, the Métis National Council, and the Indigenous Climate Action Association). The 63-member main delegation included seven indigenous representatives.[49]

With a total of 267 members, Canada's delegation at COP29 was smaller than the one that went to COP28 in Dubai the previous year (489 members for the main delegation and 149 members for the Canadian pavilion delegation, for a total of 638 members).[50] However, companies, particularly those in the fossil fuel sector, seemed to be better represented.

1.4.4. A Constellation of Observers

In addition to the official national delegations of the parties to the UNFCCC, paragraph 6 of Article 7 of the Convention stipulates that certain organizations may also attend COPs as

49. Government of Canada, "UN Conference on Climate Change: COP29 in Azerbaijan."
50. Government of Canada, "UN Conference on Climate Change: COP28 in United Arab Emirates," accessed on May 27, 2024, https://www.canada.ca/en/services/environment/weather/climatechange/canada-international-action/un-climate-change-conference/cop28-summit.html.

observers, if they so request from the Convention Secretariat. They may then be represented by their members or observers. These include the UN, its agencies, and the International Atomic Energy Agency, as well as international and national organizations, governmental and non-governmental, with expertise in the field of climate change. There are therefore three categories of observer organizations: the United Nations system and its specialized agencies, intergovernmental organizations (IGOs), and non-governmental organizations (NGOs).[51]

With several thousand NGOs admitted, they were invited—but not required to attend the COPs—to group together into nine thematic collectives known as "constituencies." These include business and industry NGOs (BINGO), environmental NGOs (ENGO), research and independent NGOs (RINGO), Indigenous Peoples organizations (IPO), the local governments and municipal authorities constituency (LGMA), the women and gender constituency (WGC), the youth constituency (YOUNGO), the trade union constituency (TUNGO), and the farmers constituency. In addition to these nine groups, other informal NGO groups exist, covering marginal themes. These include, for example, the faith-based organization group and the parliamentarian group.

1.4.5. The COPs: A Forum for Meetings, Exchanges, and Highlighting Specific Climate Change Issues

The COP *Rules of Procedure* provide that, at the invitation of the COP president, observer organizations may participate in the debates and deliberations, without the right to vote.[52] This

---

51. See: UNFCCC, *Observer Handbook for COP 29* (UNFCCC, 2024), 3, https://unfccc.int/sites/default/files/resource/Observer%20Hand book%20for%20COP29%203008%20pub%20%281%29.pdf.
52. UNFCCC, *Questions d'organisation : Adoption du règlement intérieur*, [1996] FCCC/CP/1996/2, art 6 para 2 and art 7 para 2, https://unfccc. int/sites/default/files/resource/02f_0.pdf.

is traditionally the case for the Global Environment Facility, for example.

Alongside the negotiations, official events called side events are organized to enable all participants, whether members of a government delegation or mere observers, to meet, exchange views, express their concerns, and propose solutions. These side events take the form of workshops, conferences, seminars, debates, and so on.

The COP site is sometimes subdivided into two zones. The first, blue, is accessible only to accredited participants who are part of an official delegation or act as observers. This is where negotiations take place, as well as side events that are part of the official program. However, other events no less important and unifying can be held in a second zone, the green one, open to the public, which also encourages exchange.

By accrediting the many NGOs and other members of civil society who can attend as observers, and the media coverage they generate, which puts the parties in the spotlight of public opinion and civil society, the COPs are more than just negotiating forums.[53] They also serve to inform public opinion and raise awareness of climate change. They help to highlight and shed light on specific aspects of these changes and the issues at stake, as well as mobilizing and directing measures in favour of the most vulnerable populations and groups.

Thus, for example, gender issues in relation to climate change, which had been absent from the UNFCCC, burst onto the negotiating agenda with the launch of the Lima Work Programme on Gender at COP20 in Lima in 2014.[54] The aim was to promote the integration and progressive

---

53. *United Nations Framework Convention on Climate Change*, art 7 para 6; art 13 para 8; *Paris Agreement*, art 16 para 8.
54. *Programme de travail de Lima relatif au genre*, UNFCCC, 20th Sess, UN Doc FCCC/CP/2014/10/1dd.3 (2014) Dec 18/Cp20 at pages 42–44, https://unfccc.int/sites/default/files/resource/docs/2014/cop20/fre/10a03f.pdf

reinforcement of gender equality in the implementation of the Convention by its bodies and parties, with a view to gender-sensitive climate policies,[55] in particular to ensure that girls and women do not suffer more than their male counterparts from global warming. The Paris Agreement, adopted the following year, incorporated these issues. COP22 in Marrakech in 2016 extended this program by three years, to 2019.[56] A Gender Action Plan was launched at COP23 in Bonn in 2017.[57] At the end of the extension, taking over from the program and its action plan, an Enhanced Lima Work Programme on Gender and its five-year Gender Action Plan was adopted at COP25 in Madrid in 2019.[58] At the end of these five years, at COP29 in Baku in 2024, this program was extended for a further 10 years and a new action plan was drawn up for 2025.[59]

55. UNFCCC, "The Enhanced Lima Work Programme on Gender," accessed March 28, 2025, https://unfccc.int/topics/gender/work streams/the-enhanced-lima-work-programme-on-gender.

56. *Questions de genre et changements climatiques*, UNFCCC, 20th Sess, UN Doc FCCC/CP/2016/10/Add.2 (2016) Dec 21/CP.22 at pages 18–21, para 6, https://unfccc.int/sites/default/files/resource/docs/2016/cop22/fre/10a02f.pdf.

57. *Mise en place d'un plan d'action en faveur de l'égalité des sexes*, UNFCCC, 23rd Sess, Un Doc FCCC/CP/2017/11/Add.1 (2017) Dec 3/CP.23 at pages 13–14, https://unfccc.int/sites/default/files/resource/docs/2017/cop23/fre/11a01f.pdf.

58. *Programme de travail renforcé de Lima relatif au genre et son plan d'action pour l'égalité des sexes*, UNFCCC, 25th Sess, Un Doc FCCC/CP/2019/13/Add.1 (2019) Dec 3/CP.25 at pages 7–19, https://unfccc.int/sites/default/files/resource/cp2019_13a01F.pdf.

59. *Gender and climate change*, UNFCCC (2024) Dec 7/CP.29 at para 11 and 13, https://unfccc.int/sites/default/files/resource/cp2024_11a01E.pdf.

1.5. Efficiency and Relevance Called Into Question

1.5.1. A Complex and Heavy Machinery

COPs are slow, cumbersome, and complex machines where decisions can only be taken by consensus, between nearly 200 countries with different realities, and sometimes divergent interests and strategies. This mode of decision-making has become the norm, since the COP has never officially adopted voting rules.[60] Consensus is a way of reaching agreement without a formal vote. In theory, therefore, it implies that any decision is taken without formal objection from any party; if a COP party objects to a decision, it cannot, in principle, be adopted. In practice, however, consensus is interpreted by the COP presidency. There is thus a political dimension to the process. There are precedents where formal objections were raised, but the decision was adopted by the presidency anyway. It depends on who is making the objection and who is supporting the text under negotiation. For example, at COP29 in Baku in 2024, the final agreement was immediately contested and treated as an "affront" as soon as it was adopted by several parties, in particular India, whose representative claimed that the Azerbaijani presidency had denied her the floor before the final approval of the text. The way in which texts are adopted is designed to try and resolve disagreements between the parties, although the absence of voting rules and the search for consensus tend to unbalance relations between states. Indeed, decision-making by a two-thirds majority of the parties present and voting, as provided for in the rules of procedure that have never been adopted by the COP, would help to transform the dynamics of decision-making. To date, preliminary discussions have taken place, often in small groups and informally, in an attempt to identify major

---

60. *Organizational Matters: Adoption of the rules of procedure*, UNFCCC, Un Doc FCCC/CP/1996/2 (1996), https://unfccc.int/sites/default/files/resource/02_0.pdf.

disagreements ahead of formal negotiations and to find avenues of compromise. The proposals are then presented to the parties, undergo successive revisions, and incorporate the parties' concerns, before being adopted if no objection is expressed by any party. This mode of decision-making is particularly slow, leading to the adoption of texts that often lack ambition. Aykut and Dayan have described them as "factories of slowness."[61] Their harsh observation is that:

> a hiatus has gradually emerged between a civilized and consensual UN governance process, which treats climate change as a circumscribed and governable problem, and a world reality marked by the relentless struggle for access to resources by an economic and financial globalization that feeds on the exploitation of fossil fuels, and by the spread of Western lifestyles (our translation).

This often leads to blockages or minimal compromises, particularly when there is a lack of consensus during last-minute negotiations. Voting on decisions is based on the universal principle of "one state, one vote," whereby every country, regardless of its size and importance, has the same vote. However, the negotiation process itself is not very egalitarian. Indeed, the richest states, capable of assembling delegations of several hundred negotiators active on all fronts at the COP, find it easier to assert their positions against less powerful states with smaller delegations. By way of comparison, it is interesting to note that the United Arab Emirates delegation numbered 3,491 members at COP28, compared with just 13 members for Tuvalu, for example. What is more, COPs often run overtime (COP29 in Baku was extended for an additional two days), and delegations from less affluent states are generally less able to afford to stay until the end of negotiations, incurring additional travel and accommodation

---

61. Stefan Aykut and Amy Dayan, *Gouverner le climat ? Vingt ans de négociations internationales* (Les Presses de Sciences Po., 2015), 124.

costs. In an attempt to correct these inequalities, the COP provides human and financial resources to the smallest delegations through the United Nations Development Programme
(UNDP). Despite this, it is difficult for them to follow the
negotiations in their entirety. It should also be noted that the
most vulnerable countries are largely dependent on the political will of the richest states to contribute to the fight against
climate change, unlike a forum such as the one on biological diversity, where the states of the Global South, with their
biological resources, can carry more weight at the negotiating table.

Finally, once a decision has been taken or an agreement
adopted, states always have the option, as a matter of sovereignty, not to ratify or to withdraw once they have done so.
By way of illustration, the United States signed but did not
ratify the Kyoto Protocol, and Canada withdrew from it in
2011 (effective 2012), despite being two major GHG emitters.
As for the Paris Agreement, just a few months after his election as president of the United States in 2016, Donald Trump
withdrew the country from the agreement. Elected in 2020,
Joe Biden reinstated it. However, when Donald Trump was
re-elected in 2024, withdrawing—once again—from the Paris
Agreement was one of his first decisions, made on the very
day of his inauguration on January 20, 2025.[62]

1.5.2. A Controversial Process

Some observers criticize the COPs for being increasingly infil-
trated by lobbies and industrialists from the fossil fuel sector, decrying their growing vulnerability to influence from
this sector. The growing number of representatives from this
sector is a potential factor of influence. At COP27 in Sharm

---

62. "Putting America First in International Environmental Agreements,"
the White House, January 20, 2025, https://www.whitehouse.gov/
presidential-actions/2025/01/putting-america-first-in-international-
environmental-agreements/.

el-Sheikh in 2022, where there were "only" 636 fossil fuel lobbyists counted by Corporate Accountability, Corporate Europe
Observatory (CEO), and Global Witness (GW), these lobbyists still outnumbered the representatives of the 10 countries
most affected by climate change, according to Germanwatch[63]
(Puerto Rico, Myanmar, Haiti, Philippines, Mozambique,
Bahamas, Bangladesh, Pakistan, Thailand, Nepal).[64]

Since then, their numbers have been growing. According
to Kick Big Polluters Out—a collective of over 450 environmental organizations around the world—accredited fossil
fuel lobbyists at COP28 numbered almost 2,500, the highest number ever.[65] They outnumbered each of the national
delegations, with the exception of the United Arab Emirates
(3,491 people). Some attend the COPs as observers, while
others join the official delegations of the parties, giving them
considerable weight and an active role in the negotiations.
At COP28, for example, France's official delegation included
six executives from Total Energies (including CEO Patrick
Pouyanné, who played an active role in the negotiations) and
representatives from Électricité de France; the Italian delegation included members from Ente nazionale idrocarburi
(ENI); and the European Union delegation included members from ENI, BP, and ExxonMobil.[66]

63. Germanwatch, *Global Climate Risk Index: The 10 Countries Most
    Affected from 2000 to 2019 (Annual Averages),* https://www.german
    watch.org/sites/germanwatch.org/files/2021-01/cri-2021_table_10_
    countries_most_affected_from_2000_to_2019.jpg.

64. Corporate Accountability, "Plus de 25 % de lobbyistes des énergies
    fossiles en plus par rapport à l'an dernier envahissent les pourparlers
    sur le climat de la COP," press release, November 10, 2022, https://
    corporateaccountability.org/media/lobbyistes-des-energies-fossiles-
    cop27/.

65. Kick Big Polluters Out, "Release: Record Number of Fossil Fuel Lob-
    byists at COP28," December 5, 2023, https://kickbigpollutersout.org/
    articles/release-record-number-fossil-fuel-lobbyists-attend-cop28.

66. Kick Big Polluters Out, "Release: Record Number of Fossil Fuel Lob-
    byists at COP28."

Another potential factor of influence of fossil fuel industries on COPs is that the last two COPs have been held in oil-producing countries, with COP29 taking place in Baku, Azerbaijan, in 2024 and COP28 in Dubai, United Arab Emirates, in 2023. The president of COP28, Sultan Ahmed Al Jaber, was also chairman and CEO of the Abu Dhabi National Oil Company (ADNOC), the leading Emirati oil company. Similarly, in Baku, the minister of the environment and natural resources, Mukhtar Babayev, was called upon to chair COP29, despite having worked for the State Oil Company of Azerbaijan for over 20 years. Two consecutive COPs have thus been chaired by people who have been senior executives of national oil companies.

Even though a gradual transition away from fossil fuels was enshrined in the final decision of COP28, the first time such a mention has appeared in a final COP decision, the United Arab Emirates, Azerbaijan, and Brazil, the countries that have or will chair COP28 (2023), COP29 (2024), and COP30 (2025), will paradoxically collectively increase their gas and oil production by a third between now and 2035, according to a September 2024 analysis by the NGO Oil Change International.[67] There is therefore a certain dichotomy between Dubai's commitments and the real will of politicians and the business world. According to a 2023 report by the same NGO, five parties to the UNFCCC—the United States, Canada, Australia, Norway, and the United Kingdom—are responsible for half of the world's projected new deposits by 2050.[68] Total Energies, whose CEO was part of the official French delegation at COP28, is the third company in the

---

67. Oil Change International, "Countries Tasked with Leading Efforts to Increase Climate Action Increasing Oil and Extraction by 33%," news release, September 25, 2024, https://www.oilchange.org/news/countries-tasked-with-leading-efforts-to-increase-climate-action-increasing-oil-and-gas-extraction-by-33/.

68. Oil Change International, *Planet Wreckers: How Countries' Oil and Gas Extraction Plans Risk Locking in Climate Chaos* (Oil Change

world (after ExxonMobil and Brazil's Petrobras) whose plans for developing hydrocarbon exploitation are the most incompatible with the objective of keeping global warming to 1.5°C, according to the NGO Urgewald.[69] The risk of conflicts of interest is therefore very real at COP meetings.

As a result, the COPs are widely criticized, both in terms of the way they operate and the decisions they adopt, notably because progress is deemed to be minimal in view of the urgency of climate change, and because these decisions are not always followed by action. They are also criticized for allowing certain governments and economic players to engage in greenwashing, by giving the illusion of taking action without seeking to bring about real change.

And yet, the UN is constantly warning of the urgency of climate change. The 1.5°C limit advocated by the Paris Agreement in 2015 has already almost been reached less than 10 years later, while we had to wait more than 30 years after the adoption of the UNFCCC in 1992 to see the adoption of a fossil fuel phase-out resolution at COP28 in 2023. UNEP's 2024 *Report on the Gap between Needs and Prospects for Reducing Greenhouse Gas Emissions* forecasts a "catastrophic" warming of 3.1°C by the end of the century if current actions continue.[70]

1.5.3. An Ambitious Role for International Climate Diplomacy?

Despite this, the role of the COPs should not be overlooked. By bringing people together, they have enabled the emergence of international climate diplomacy involving virtually every state on the planet. They have also enabled the establishment

International, 2023), https://www.oilchange.org/wp-content/uploads/2023/09/OCI-Planet-Wreckers-Report.pdf.

69. Urgewald, "Global Oil & Gas Exit List." See: Data Highlight: Expansion, Table 1.5°C Incompatible Development Plans, https://gogel.org/.

70. UNEP, *Emissions Gap Report 2024: No More Hot Air … Please!* (UNEP, 2024), https://www.unep.org/resources/emissions-gap-report-2024.

of an international legal architecture that guides, frames, and legitimizes national policies and legislation, which many states would not have initiated and implemented without them.

In fact, the myriad of COP decisions, clarifying and operationalizing the UNFCCC, have enabled the development of common frames of reference on an international scale. In particular, the adoption at the COPs of major multilateral agreements such as the Kyoto Protocol and the Paris Agreement represents a major step forward. Together with the UNFCCC, these agreements form the backbone of modern environmental law at both international and national level and are subsequently incorporated into national law.

As Sandrine Maljean Dubois points out, such agreements "represent the most effective tool for inter-state cooperation to date." They have made the notion of sustainable development an essential part of public policy. They have defined clear global objectives—such as the Paris Agreement's ambitious target of limiting global warming to +1.5°C—on which national actions can be based, and which the most ambitious policies can aspire to exceed. Indeed, nothing prevents national governments from going further in the fight against global warming, if they wish or are able to do so.

Be that as it may, the current COP climate system seems a priori to be the subject of debate as to its strengths and weaknesses, its track record, and its future. By sharing experiences of participation over its 30 years of existence, the present book aims to enrich this debate by putting forward the diversity of viewpoints and perspectives of people who, over the years, have seen the COP come into being and evolve. Offering a privileged glimpse behind the scenes of the COP and a clear and accessible reading of its complex and fascinating—not to say nebulous—mechanisms, the contributions gathered in this book will also enable everyone to better understand the workings of the COP climate system to better appreciate its ins and outs. In so doing, the book aims to equip every Canadian with the tools to participate in the public debate on the future of our planet.

CHAPTER 2

History of the COP
from COP1 to COP30

Jean Lemire

2.1. Scientific Evidence

Climate change is one of the greatest challenges humanity faces.[1] We know it. Science backs it up. The United Nations

---

1. In this chapter, I offer a personal analysis of the major international climate meetings. My views and comments are inspired by my experiences as an observer of the international environmental scene, which date back to the early 2000s, when I sailed around the world to document the impact of climate change. Over my 30-year career as an explorer and filmmaker, I took an interest in humanity's impact on the planet's major ecosystems. After several expeditions to the Arctic and Antarctic, I realized just how urgently we needed to take action and set out to document the shifting relationship between humans and nature. For nearly a decade, I have held the position of envoy for climate change, northern and Arctic affairs for the Government of Quebec. I also take part in the COPs as a member of the Canadian delegation and represent the federated states within the UN Convention on Biological Diversity. This has given me the ability to share my observations and personal analysis as a privileged observer of the COPs. I am neither a lawyer nor a historian; readers will have to forgive several legal and historical shortcuts. Rather, I must concede and acknowledge that this document is merely a record of my understanding and vision. My words represent only my personal opinions and should not be interpreted as official positions of the governments of

declares it loud and clear. Books, films, and other media reiterate what 97 percent of the world's scientists have been saying for a long time: Humans are responsible for an unprecedented crisis of their own making.

And yet, after more than two decades of multilateral negotiations, we still have not managed to reduce the greenhouse gases (GHGs) that threaten our very survival. Humanity has failed miserably in its commitment to correct a situation that economic models have created.

The climate crisis should have our governments declaring an immediate state of emergency, ordering a complete overhaul of the way we consume and finance development, and reconceptualizing humans' place within planetary ecosystems that are under repeated attack from our gluttony as a species. Scientists speak of a new geological era, the Anthropocene, characterized by the influence of humanity as the main driver of change on Earth, overpowering geophysical forces and a millennia-old balance. That is no mean feat!

Humans and the planet have always coexisted. Yet today, this relationship is crumbling. For the first time in Earth's history, human activity has become one of the main drivers of global change. The climate has always been central to our great social and economic movements. The rapid decline in ecosystem functions and services has upset an equilibrium that has lasted for millennia, especially the ability to regulate the climate over the long term. Biodiversity is also a victim of the age of humanity, to the point where scientists now fear a sixth mass extinction. Pollution, whether it comes from our atmospheric emissions, from our waste management that spreads microplastics over every surface on Earth,

---

Quebec or Canada. Responsibility for my writings cannot be claimed by, or granted or arbitrated to, any other entity, employer, or organization. As I wrote, I used journalistic accounts and references to clarify dates, to document several events where I was absent, and even to borrow certain quotes. However, I have deliberately not provided full citations, which would have weighed down the text.

from our unsustainable agricultural practices that drench the land in chemical fertilizers and pesticides, or from the shameless punishment we inflict on our streams, rivers, lakes, and oceans, is now directly impacting our health.

In its sixth report, the Intergovernmental Panel on Climate Change (IPCC) also revealed that climate risks are rapidly increasing in frequency and intensity, and that 3.6 billion people already live in areas that are highly vulnerable to climate change. Low-income countries and Small Island Developing States suffer the most serious health repercussions, even though their contributions to global GHG emissions remain marginal. In vulnerable regions, the mortality rate of extreme weather events over the past decade has been 15 times higher than that of less vulnerable regions. The World Health Organization (WHO) has demonstrated that:

> Climate change is impacting health in a myriad of ways, including by leading to death and illness from increasingly frequent extreme weather events, such as heatwaves, storms and floods, the disruption of food systems, increases in zoonoses and food-, water- and vector-borne diseases, and mental health issues. Furthermore, climate change is undermining many of the social determinants for good health, such as livelihoods, equality and access to health care and social support structures. These climate-sensitive health risks are disproportionately felt by the most vulnerable and disadvantaged, including women, children, ethnic minorities, poor communities, migrants or displaced persons, older populations, and those with underlying health conditions.

And if nothing is done soon, the WHO warns that "[f]urther delay in tackling climate change will increase health risks, undermine decades of improvements in global health, and contravene our collective commitments to ensure the human right to health for all."

One might think that the clear, scientifically proven relationship between the environment and health would accelerate multilateral commitments. Evidence of the growing threat continues to mount. Given the scientific proof, not to mention the catastrophic human toll of our out-of-control relationship with the planet, humanity has a greater duty than ever to anticipate, manage, and take action to ensure historic resilience in the face of the threat to its very existence.

It is often said that current multilateralism is in crisis. People criticize the slowness and lack of ambition in decision-making, rightly point out bureaucratic paralysis, and even accuse international organizations of bowing to outside pressure. But multilateralism is an ambitious normative project that seeks to establish an inclusive and cooperative world order. Such a mission is bound to bring its share of challenges. Throughout history, multilateralism has regularly had to contend with power politics, the unilateralist approach of certain countries in a position of strength, and even major conflicts that have called its very existence into question. However, this form of global cooperation remains unique and resilient. Historically, the network of international organizations (like the UN, the World Bank, and the WHO) has survived; it continues to coordinate joint diplomatic action, which is, by definition, humanity's response to collective threats. So why are we not rising to the environmental challenge when our very survival is at stake? Why are the COPs—international gatherings that can draw over a hundred thousand participants a year—unable to come up with a collective, concerted defence against the greatest common enemy that nations have ever faced? To try and understand the current reluctance to counter major environmental and social challenges, we need to look back and examine the contentious, complex relationship between the environment and the economic development of our societies.

2.2. The Beginnings of Environmental Diplomacy

The first United Nations Conference on the Human Environment was held in Stockholm, Sweden, in 1972. This international meeting highlighted the role of the environment in the global action plan and led to the creation of UNEP, the United Nations Environment Programme. For the first time, representatives of a certain political class recognized the fundamental link between the environment and societies' economic development. The interdependence of humans and the Earth was finally acknowledged, and most people (barring some skeptics) became aware of the already-damning scientific prognoses for the threat humanity poses to the planet's ecosystems. The world was already showing signs of concern, and it became clear that we needed to unite to coordinate international action.

Public policy had to rise to the challenge. It could no longer be limited to states and nations; the world needed a global approach to the issue. Environmental disruption was understood as a common enemy, one that called for quick, effective, multilateral action. International meetings proliferated over the next two decades, resulting in reports and international agreements that placed environmental issues at the heart of what would eventually be known as "sustainable development." This meant reviewing our economic models to make sure that development would not overtax resources and ecosystems, and that it would meet current needs without compromising the ability of future generations to do the same. This idea of sustainable development introduced a conciliatory process that focused on economic progress, social progress, and implicit recognition of future generations. In theory, it was a promising concept for humanity and a call for concerted, global action. In a way, we were asking countries to act for the common good. Sustainable development, as it was defined, therefore suggested setting limits on all-out economic development. It called on scientific evidence to emphasize the importance of guiding economies with a vision that ensures the sustainability of ecosystems. What is more, it insisted on social justice as a fundamental aspect of economic development. We set ourselves the task of combating the environmental disturbances created by our economies of extraction and consumption, notably by regulating certain economic activities, while addressing inequalities. In this era of change and consideration for the future, we hoped to take advantage of the proposed environmental reforms to revisit the entire relationship between the Global North and Global South, and thereby tackle social and economic inequalities. At the time, for multilateralism, this was a very broad concept, but the scientists' findings were incontrovertible. We needed to act, on all fronts.

In 1992, on the 20th anniversary of the Stockholm Conference on the Human Environment, countries organized and invited the world to the first United Nations Conference on Environment and Development (UNCED), also known as the Earth Summit, in Rio, Brazil. Representatives from 178 countries, as well as non-governmental organizations (NGOs) and civil society, met in what would become the forefather of today's Conferences of the Parties (COPs). Nearly 30,000 people came together to discuss, negotiate, and debate environmental and sustainable development issues. They covered a multitude of environmental topics, including the protection of land, air, and water; the conservation of biological diversity, forests, and natural resources; desertification from soil degradation; and ozone depletion and global warming, to name but a few.

Other issues were covered as well. Poverty in the developing world, economic growth, and consumption patterns, which widened the gap between developed and developing countries, were all discussed in Rio. Participants also debated the tensions between rich and poor, the colonialist approach often associated with the exploitation of natural resources, and historical responsibility for the state of the planet. Finally, in view of the urgent need for action—and despite major differences of opinion on the financial resources to be invested—a set of mechanisms were adopted to implement the findings from the Earth Summit. In particular, this major international meeting gave rise to three conventions: the (UNFCCC), the United Nations Convention on Biological Diversity (UNCBD), and the United Nations Convention to Combat Desertification (UNCCD).

2.3. Common but Differentiated Responsibilities

As early as 1992, we agreed on a fundamental principle: Global disruption is not the result of a collective, unified, and equal effort by humankind. Since the industrial age, environmental damage has largely stemmed from the economic development of a minority. Historically, developing and highly vulnerable countries have played a relatively small part in chipping away at biodiversity and causing climate change and desertification when compared to developed countries. Given that the consequences of the damage to our planet stem largely from the economic activities of a relative few, it is easy to understand how the majority can view themselves as collateral victims of this minority and its efforts to become richer. What is more, the countries that are hardest hit by this environmental damage are precisely those that have contributed to it the least.

In light of this realization, the developed countries at the Earth Summit recognized the part they needed to play in the international effort. Given the pressures that wealthy societies have long exerted on the environment, the principle of "common but differentiated responsibilities" became enshrined in international environmental conventions. This provides a legal basis for determining how obligations differ proportionately to the states' responsibilities and indirectly ensures a certain balance in their relations regardless of economic, political, or social disparities. Recognition of developed countries' historical responsibility for environmental damage has led to some progress in multilateralism. However, the language in Article 3.1 of the UNFCCC that describes the contribution of developed countries has changed over time, gradually weakening the legal obligation and even calling the notion of equity between states into question.

2.4. Conferences of the Parties and the Kyoto Protocol

The first COPs, which started in 1995, enabled member countries to work together to set trajectories for GHG emissions and attempt to put the planet back on the path to sustainable development. The first global climate agreement, known as the Kyoto Protocol, was signed at COP3 in 1997. However, the treaty's ambitions fell far short of scientists' demands. Some countries used the consensus rule, voted on at the very first COP in Berlin in 1995, to slow down, block, or weaken the agreement. The idea of a two-thirds majority vote has always been rejected by a handful of countries which do not want to lose their power to influence—or even block—majority decisions. This principle of consensus for COP decisions is still debated and debatable today. It explains in part why UN processes are so slow and cumbersome, and why there seems to be a lack of ambition when it comes to tackling such massive environmental challenges. The consensus requirement acts as a moral pressure on countries and their leaders, but it becomes a major obstacle to negotiations when bad faith, whether political or economic, prevails over the desperate need for ambitious commitments. Some countries are very adept at using this rule of law to paralyze negotiations.

The 1997 Kyoto Protocol set binding GHG reduction targets for industrialized countries but did not oblige developing countries to reduce their emissions. More specifically, the Protocol called for industrialized countries to reduce their GHG emissions by at least 5.2% (compared to 1990 levels) by 2012. For the first time, governments were adopting quantified, legally binding reduction targets. A compliance monitoring committee even permitted sanctions if countries did not comply.

During negotiations, the principle of common but differentiated responsibilities was brought up again so that industrialized countries' commitments could be scaled based on the level of development. While international law guarantees a certain number of rights and obligations to all states, the Kyoto Protocol applied the principle of differentiated responsibilities based on countries' historical contributions since the Industrial Revolution. For example, the reduction target was set at 6 percent for Canada and Japan, 8 percent for the European Union, and 7 percent for the United States, the largest GHG emitter at the time.

The signatories were required to ratify the Protocol to make it legally binding under their own domestic laws. Additionally, it could only be implemented once 55 developed countries, representing at least 55 percent of the GHG emissions covered by the agreement, had ratified it. On March 13, 2001, American President George W. Bush announced his refusal to ratify the Kyoto Protocol, claiming that the American way of life was sacred. At the time, he argued that the Kyoto Protocol could harm the American economy. He set the condition that the international treaty should also apply to certain so-called developing countries, excluding only the poorest ones. The American approach was intended to include countries such as China, India, and Brazil, which were experiencing major economic growth at the time.

In multilateral systems, the ambitions of states are often linked to those of the political leaders in power. In the United States, the Republican president asserted his opposition to the Kyoto Protocol, citing his disbelief at the fact that certain countries were not included: "I oppose the Kyoto Protocol because it exempts 80 percent of the world, including major population centers such as China and India, from compliance [...]." His decision represented a historic uppercut, not only to the first true international treaty on climate diplomacy, but also to the fundamental principle of common but differentiated responsibilities.

The withdrawal of the United States delayed the Protocol's entry into force. It did not become legally binding until 2005, but the legal obligation to meet GHG reduction targets had some leaders questioning the value of an international treaty that no longer included the world's largest GHG emitter.

In 2006, Canada elected a Conservative government. Canada's new prime minister, Stephen Harper, had already expressed his views on the Kyoto Protocol in a letter sent to his party members in 2002: "Kyoto is essentially a socialist scheme to suck money out of wealth-producing nations." In the same letter, he asserted that implementation of the treaty would greatly harm the country's hydrocarbon industry, essentially echoing President Bush's argument about the consequences for local economies in industrialized countries. What is more, he took the liberty of describing the scientific evidence for climate change as "tentative and contradictory," claiming in the same breath that carbon dioxide is "essential to life."

Once in power, however, Prime Minister Harper had to reconcile his positions with the environmental values of some Canadians, who were calling for a strong, clear Canadian commitment to the environment. His government was a minority one, and Canada was divided between oil- and gas-producing provinces and others, whose economies did not directly depend on fossil fuels. He was re-elected in 2008, but once again failed to win a majority government. Under the Conservative government, Canada continued to attend the COPs, but only half-heartedly. The country was accused of systematically obstructing negotiations. Steven Guilbeault, a young environmentalist at the time, who would later become Canada's minister of the environment under a Liberal government, regularly expressed his disapproval of the Canadian attitude: "It's heartbreaking to see the bad faith with which the Conservative government has participated in recent international climate talks, always aiming to derail a renewed Kyoto Protocol."

The roadmap drawn up at COP13 in 2007 called for the negotiation of a new international climate agreement to replace Kyoto, whose first commitment period was ending in 2012. GHG reduction targets had to be revised as emissions soared. Above all, it was crucial to listen to scientists' pleas and attempt to limit global warming. This meant that there was a certain urgency in the preparations for COP15, which was to be held in Copenhagen, Denmark, in December 2009. Even back then, some people were already calling it the Last Chance Summit.

The Copenhagen Conference brought together 115 world leaders, making it one of the largest multilateral events in the history of climate diplomacy. Barack Obama had become president of the United States, and the dialogue he had begun with China promised major advances. Over 45,000 people were accredited for the event, and there was real hope that an ambitious international accord could be reached. However, the Danish presidency quickly drew criticism for its lack of transparency. Negotiations were taking place in hallways and hotel rooms with a limited number of countries. There were numerous procedural incidents, and the rules of international negotiation were not followed. Developing countries wanted to ensure that the commitments were legally binding, but to do so, they needed to take part in the negotiations, which remained restricted. The vast majority of NGOs and observers were also excluded from the negotiations. Tempers started to rise and COP15 appeared to be heading towards failure.

Tensions were also mounting between the world's two biggest GHG contributors at the time, China and the United States, who were negotiating behind closed doors. China was engaging in empty-chair politics and given the rule of unanimity during COPs, the Copenhagen Summit was headed for failure. President Obama was trying to convince Chinese

Premier Wen Jiabao. Negotiations were taking place without inviting Europeans or even developing countries.

China was vehemently opposed to any verification that the accord was being implemented. The Americans, on the other hand, doubted the honesty of China's commitments and wanted to be able to verify Beijing's CO_2 reduction results. The conflict persisted and, after a series of incredible twists and turns, a three-page agreement was finally reached, drafted in a closed room by five states (China, the United States, Brazil, India, and South Africa). The Copenhagen Accord, which was never adopted in a plenary session, has been harshly criticized by the international community. For environmentalists and representatives of civil society, it was a dismal failure.

The accord allowed China and the United States to preserve their sovereignty in deciding which national targets to adopt, and to avoid legally binding commitments.

"We will honour our word with real action," assured Chinese Premier Wen Jiabao. Barack Obama explained his position by criticizing the binding nature of the existing treaty: "Kyoto was legally binding and everybody still fell short anyway."

This marked a turning point in the history of climate diplomacy. Reduction targets would henceforth be set by individual countries, and the idea of a legal obligation had been consigned to the dustbin of history. From that point forward, climate efforts would be unilateral and non-binding. With Kyoto set to expire in 2012, the Copenhagen Summit was intended to pave the way for a new, legally binding global climate agreement. History will remember above all the failure of this international meeting, which had mobilized an impressive number of heads of state.

The following year, at COP16 in Cancún, Mexico, some of the main principles set out in the Copenhagen Accord were adopted. Agreements were reached to limit global warming to 2°C and the creation of the Green Climate Fund was announced to finance developing countries' efforts to mit-igate and adapt to the inevitable consequences of climate change. More than 130 countries accepted the agreements, including Canada, which committed to reducing its GHG emissions by 17 percent compared to the 2005 baseline year. Canada was criticized for its lack of ambition, since its new target equated to a reduction of only 3 percent compared to the 1990 baseline year, a far cry from the commitments made under the Kyoto Protocol.

At this point, Canada was still governed by the Conservatives. After two minority terms, they finally won a majority in the House of Commons in 2011, giving them free rein to act against the Kyoto Protocol, which had been ratified by Canada. In 2012, just as the Kyoto Protocol's first period was set to expire, the Conservative government repealed Canada's Kyoto Protocol Implementation Act—the first Canadian law to set mandatory GHG reduction targets—and officially withdrew Canada from the international treaty. It was a first in the history of climate negotiations.

Legally, Canada was entitled to do this under Article 27 of the Protocol, which allowed any country to withdraw after three years of its entry into force. In 1997, Canada had com-mitted to reducing its emissions by 6% below 1990 levels. Yet in 2009, its emissions had risen by 17%. Development of the oil sands in western Canada, which require a highly polluting extraction and processing process, had boomed as oil prices rose, contributing to Canada's newfound fortune as an oil-producing country. Theoretically, the legally bind-ing nature of the Kyoto Protocol targets meant that Canada would have had to avail itself of one of the Protocol's flexi-bility measures and purchase carbon credits. The minister of the environment at the time, Peter Kent, put the figure at $14 billion. This justification was based on a false premise and was deemed legally insufficient. Canada's withdrawal two weeks before the target deadline was, in effect, a violation of its international obligations. It also marked an ideological shift driven by the party in power, which felt it could do as it pleased now that it had a majority government.

Canada's own interpretation was that it was not accountable for its failure to meet its targets, since it had repealed the law obliging it to do so. However, many legal experts found this decision questionable, since the withdrawal should only have taken effect a year after notice was given. What is more, withdrawal from an international agreement is not valid retroactively, and cannot be used to justify a country's failure to meet obligations before it officially withdraws. Canada's decision set a historic precedent that weakened the treaty and once again called into question the binding nature of targets adopted as part of a multilateral process.

Canada's attitude shook the world of international negotiations, a multilateral mechanism that was already floundering. This historic withdrawal tarnished the image and reputation of Canada, which had been one of the key states in the organization of the Rio Earth Summit. The executive secretary of the United Nations Framework Convention on Climate Change at the time, Christiana Figueres, reacted:

> I regret that Canada has announced it will withdraw and am surprised over its timing. Whether or not Canada is a Party to the Kyoto Protocol, it has a legal obligation under the Convention to reduce its emissions, and a moral obligation to itself and future generations to lead in the global effort. [...] Industrialized countries, whose emissions have risen significantly since 1990, as is the case for Canada, remain in a weaker position to call on developing countries to limit their emissions.

Canada had effectively chosen to align itself with the United States, which refused to ratify a legally binding international treaty. The Copenhagen Summit had, in a way, rejected the binding nature of the targets the countries had adopted for a future agreement, thus giving Canada new arguments to justify its inaction. By announcing its withdrawal at the Durban Summit in 2012, the Canadian government torpedoed the Kyoto Protocol and sent a message to the rest of the world that it was publicly turning its back on the Protocol as an international tool for combating climate change. Canada announced that it would set its own GHG reduction targets, officially opening the door to a non-binding reduction process. Canada's decision was a clear demonstration that no country could be guaranteed to meet its GHG reduction commitments, once again casting doubt on the legally binding nature of the Kyoto Protocol and, above all, of any future agreement to replace it.

Despite the setbacks, the international community continued to meet annually for the COPs in the hopes of developing a successor to the dying Kyoto Protocol. Admittedly, the Kyoto concept was quite binary. On one side, there were the rich industrialized countries who were responsible for the climate crisis, and on the other, the developing countries, the first victims of climate change.

The Kyoto Protocol was a product of its time. China and India were not yet considered emerging economies when it was implemented in the early 1990s; at that point, China accounted for just 10% of the planet's total emissions. Yet by Canada's withdrawal in 2012, China was emitting at least 27% of global GHGs (although Chinese numbers were difficult to determine with much accuracy at the time).

The Kyoto Protocol had been founded on the principle of common but differentiated responsibilities. However, given that the United States never ratified the Protocol and that Canada was able to withdraw without legal consequences, the very notion of historical responsibility was being ignored. What is more, with these two countries not participating, Kyoto covered barely a third of the world's emissions. Its successor needed to cover all the planet's major polluters in order to obtain a new commitment from all states.

The COPs in Durban (2011), Doha (2012), Warsaw (2013), and Lima (2014) focused on developing a roadmap for the major Paris conference in 2015. Several countries were insisting that the future agreement be legally binding, but negotiations were being stretched to the limit, as was the very principle of international law on the obligations of countries.

2.5. The Paris Agreement

In 2015, countries met in Paris at COP21 to negotiate what would be humanity's new response to climate change: the Paris Agreement, which was adopted unanimously by all 196 Parties.

The Agreement did away with Kyoto's two-track system, which only required developed countries to reduce their CO_2 emissions and recognized that climate change was a shared challenge. As such, it called on all countries to set emissions reduction targets. The principle of common but differentiated responsibilities was still recognized, but applied very differently. Under the Paris Agreement, all countries need to submit nationally determined contributions (NDCs), which represent their plans to reduce their GHG emissions at the national level. These NDCs apply to all countries, regardless of their level of development, and have to be revised upwards every five years.

The legal scope of the new Paris Agreement lies in its nuances. Strictly speaking, national GHG reduction commitments (NDCs) are not binding targets, since they are specific to each country and the level of ambition is voluntary. Furthermore, there is no legal requirement to adhere to the NDCs, as they are not, strictly speaking, an integral part of the Paris Agreement. Articles 3 and 4 of the Agreement do oblige each country to set a reduction target, implement it, and, above all, revise it upwards every five years. However, the Agreement does not include any coercive mechanisms or sanctions for countries that fail to meet their commitments, making it difficult—if not almost impossible—to challenge them under international law. Normally, the main recourse would be the 1969 Vienna Convention on the Law of Treaties, which requires treaties to be "performed by [the parties] in good faith." But the Paris Agreement is not officially called a treaty or a protocol, another important legal nuance. If it were, it would have needed to be ratified by the US Congress, which was predominantly Republican and climate skeptic at the time. To avoid a filibuster in the Senate, John Kerry, the US secretary of state representing the Obama administration, intended to ratify the agreement through an executive order. However, the terminology in the Agreement had to be scrutinized in order to bypass Congress.

On the eve of the final plenary session, as everyone waited patiently for the outcome of the last rounds of negotiations, murmurs filled the large meeting room, which was so packed with accredited observers that other rooms had to be opened with live broadcasts. I was there, along with several representatives of civil society and NGOs from all over the world. All were genuinely hopeful of reaching an ambitious international agreement. The wait was unusually long, foreshadowing last-minute disagreements. Observers were frantically checking their phones, hoping to get inside information on what was being discussed behind closed doors. Among them was Steven Guilbeault, a well-known Canadian environmentalist. He had taken part in all the COPs and knew everyone there. He had a knack for gathering information from hallway conversations, which made it easier to follow the behind-the-scenes negotiations we could not see. He quietly informed me that there were rumours that the United States was refusing to accept certain words in the final text, whereas China was pressing for no other changes to be accepted. As this type of multilateral negotiation requires unanimity, in principle the text had to be reopened for comments by the parties if any country proposed even minor changes.

In the final version of the Agreement, the US administration had discovered a five-letter word that could have changed everything. In the fourth paragraph of Article 4, the word "should" had been replaced by "shall," covertly creating a legal obligation towards developed countries. More specifically, the phrase "[d]eveloped country Parties should continue taking the lead by undertaking economy-wide absolute emission reduction targets" now read, "[d]eveloped country Parties shall continue taking the lead [...]." The word "shall" created an obligation, while the word "should" was more of a non-legal suggestion (even today, some legal experts question this difference between "hard" and "soft" law). US Secretary of State John Kerry was adamant: "We cannot do this and we will not do this. Either [the term] changes or President Obama and the United States will not be able to support this agreement."

There was a fine line between what could be written and what would become legally binding, and they were doing everything in their power to avoid a vote in Congress. For some countries, the replacement of the word justified a reopening of negotiations; this had to be avoided at all costs given that deadlines had already been pushed back.

It took interventions at the highest level between heads of state to agree to change the little five-letter word and thus prevent the Paris Agreement from failing. All that remained was the final step: adoption by consensus in a plenary session. The COP21 presidency was nervous. The last amendment, which removed an important binding legal aspect, had been negotiated behind closed doors and without including all the parties.

Finally, much later than announced on the evening of December 12, 2015, the COP21 president, French Foreign Minister Laurent Fabius, surprised everyone by acting very swiftly on the podium. He had no intent to reopen the text of the Agreement for what was deemed a typographical error. And yet, that little five-letter word, which made the Agreement legally binding, was what many countries wanted. After a short onstage introduction, he quickly went over the negotiations, gave a cursory overview of the Agreement, then looked out at the room full of delegates and said, "I hear no objection. [Very brief pause] The Paris Agreement is accepted." And with the bang of a gavel, he officially marked the end of a years-long negotiation process.

The speed with which the French minister acted surprised many, including the interpreters who were unable to relay the information in the UN's five official languages in time. It took the delegates and observers a few seconds to realize what was happening. Laurent Fabius and the other close negotiators of the Agreement all stood up at once on the podium to celebrate the moment. The delegates and representatives of civil society were taken by surprise, but largely reacted with jubilation. It had not quite sunk in, but the first-ever global climate agreement had been adopted and all countries were taking part! After two weeks of intense negotiations in Le Bourget on the outskirts of Paris, the recalcitrant countries had finally rallied behind the historic agreement.

The French presidency of COP21, under Laurent Fabius, had spared no effort, stretching informal meetings well into the night. There had been serious doubts about multilateralism since the Copenhagen failure, and people were cautiously hopeful at best. In the end, the Paris Agreement brought together the United States and China, the two biggest polluters on the planet, as well as India and oil-producing countries like Venezuela and Saudi Arabia, which were known for opposing any treaty that directly targeted the fossil fuel industry.

This was a great and historic victory, marked by optimism and renewed ambition from certain countries—including Canada that had just elected a Liberal government, which, in stark contrast to its unambitious predecessor, arrived in Paris with a five-principle plan to address the major climate ambitions the UN had been demanding for decades. Canada's return as a leader in the fight against climate change was welcomed and celebrated by the international community. Canada's new prime minister, Justin Trudeau, uttered his trademark phrase to the heads of state gathered in Paris: "Canada is back, my good friends. We're here to help." With its minister of the environment, Catherine McKenna, Canada played an important role in adding the 1.5°C threshold by backing several countries, including Small Island Developing States, that had requested the change. There is no question that, in Paris, Canada presented itself as a country resolutely committed to fighting climate change. However, while the words were encouraging, the new government had yet to prove that it had taken concrete action.

2.5.1. The Objectives of the Paris Agreement

The Paris Agreement has three key objectives: mitigation, adaptation, and finance. For the first step, countries committed to keeping global warming to below 2°C above pre-industrial levels and continuing efforts to limit the increase to 1.5°C. The agreement also sought to cap GHG emissions as quickly as possible in order to reach a balance between the quantities of GHGs emitted and those captured by natural carbon sinks and future CO_2 capture technologies. The idea of achieving carbon neutrality by 2050 was formalized for the first time.

The countries also committed to strengthening their capacities to cope with the impacts of climate change. With projections warning of rising global temperatures, not to mention the scientific reports that show the potentially catastrophic impacts of climate change, humanity had no choice but to develop programs to adapt to climate disruption. This resolution targeted developing countries and Small Island Developing States in particular.

Finally, the countries agreed to invest financial resources that are compatible with the adopted climate objectives, starting with the mobilization of $100 billion per year for developing countries as of 2020.

Ultimately, the Agreement sets a global trajectory, but recognizes the differing circumstances of the countries as they determine their own contributions to that trajectory. It also acknowledges the key role played by non-state actors, starting with subnational governments, cities, and regions, but not forgetting members of civil society, like Indigenous communities, youth, women, academics, and businesses. This inclusive, participatory approach submits the challenge of climate change to the whole-of-government and whole-of-society approaches.

As there are no coercive measures to compel countries to meet their NDC commitments, the Paris Agreement includes an Enhanced Transparency Framework. Under this framework, each country is required to report transparently on the measures taken and progress made in terms of climate change mitigation, adaptation, and financing (given or received). The reports measure countries' efforts and evaluate collective progress towards the long-term global objectives set by the Agreement. This Global Stocktake is used to formulate recommendations to guide the ambitions of countries, which must revise their targets upwards every five years.

This transparency mechanism (Article 13, which covers monitoring, reporting, and verification—or MRV—measures) applies to all countries, including developing ones, albeit with greater flexibility. The transparency required by the Paris Agreement is built on trust and dialogue between countries, with the aim of global cooperation to align ambitions with the Agreement's targets. It relies on a "name and shame" system to publicly denounce the states that are not meeting their targets. Accountability for reduction targets at the national level therefore puts a country's reputation on the line, not only in the eyes of other signatories to the Agreement, but also in the eyes of civil society, which can use the disclosed data to exert pressure by denouncing the country's failure to meet its commitments. It is more of a moral sanction than a legal obligation, although some legal experts consider the Paris Agreement to be legally binding (which is how it is presented on the UN website).

The adoption of the Paris Agreement by acclamation was just the first step in the UN process. Countries still needed to ratify the agreement at the national level, a process that can take years. But its adoption with such fanfare in Paris spurred sustained international momentum. The UN secretary-general at the time, Ban Ki-moon, invited countries to New York for the official signing of the text on April 22, 2016. On that Earth Day, government officials and heads of state from 160 countries travelled to the United Nations headquarters to take part in the event. It was a strong symbol intended to spur enough countries to ratify the Agreement at the national level for it to be formally implemented. Under international rules, for the Agreement to come into force, at least 55 percent of countries accounting for at least 55perecnt of global emissions had to ratify it.

Discussions continued between American President Barack Obama and Chinese President Xi Jinping, the leaders of the two largest GHG emitters. At the time, the two countries accounted for 38 percent of global emissions. As expected, the Democratic president proceeded by executive action to avoid submitting the Agreement to a Republican-majority Congress. In China, the president controls the National People's Congress. Both countries quickly ratified the Agreement (September 3, 2016), sending a strong message to the rest of the planet. The Paris Agreement entered into force in record time, on November 4, 2016. In all, 96 countries had already ratified it, marking the return of real global determination to fight climate change.

The United States' hurry to ratify the Paris Agreement was understandable. President Obama was at the end of his term and the Republican candidate for the US presidential election was a climate change skeptic, wealthy businessman, and reality TV show host who had never held public office. Donald Trump was an atypical and unprecedented presidential candidate in more ways than one.

Despite an upturn in the US economy, the election themes centred on the precariousness of the working classes and the war on terrorism. The Republican candidate ran a disorderly and aggressive campaign that flouted the most fundamental rules, starting with those surrounding truth and facts, which he often disregarded in his lengthy speeches.

Donald Trump is an outspoken isolationist. His statements and intentions during the campaign proposed a unilateralist approach for the United States creating a slogan within and alongside the Republican Party: "Make America Great Again" (MAGA). According to the polls at the time, it was unlikely that he would be elected president. His opponent, Democrat Hillary Clinton, was leading in the majority of swing states. But Trump represented a different political path. He was arrogant and often vulgar, but above all, he was focused almost exclusively on the United States. In his speeches, he did not hesitate to criticize multilateralism and international cooperation, criticism that was unprecedented in American political discourse. Trump has proved to be incredibly resilient, despite the numerous scandals that have come to light and which would normally be disqualifying for any other candidate. He succeeded in shaking the foundations of democracy by speaking directly to a segment of the American population that rejected the American establishment. As a former first lady, Clinton represented that establishment class in the eyes of many Americans. Despite being a billionaire raised in luxury, Trump still managed to speak to the people. His misleading rhetoric appealed to a large proportion of Americans who saw him as a kind of saviour, an anti-establishment candidate who would bring change. Several party members, despite being leading Republican figures, actually distanced themselves from their candidate during the campaign due to the extreme nature of his racist, misogynistic, and anti-democratic positions. But nothing seemed to affect or faze the atypical candidate.

On November 8, 2016, four days after the Paris Agreement came into force, there was a momentous upheaval. In a stunning turn of events, the United States elected Republican Donald Trump, the climate change skeptic, to serve as their 45th president. On June 1, 2017, in keeping with his election promise, he proclaimed that the United States would withdraw from the Paris Agreement. The announcement was a bombshell to the diplomatic world and was roundly criticized by the vast majority of countries. According to the legal rules of the Agreement, the withdrawal would not come into effect until 2020; the text negotiated by former President Obama stated that no country could leave the Agreement before the third anniversary of its entry into force, and countries that signal their intent to withdraw still have to wait a year after providing official notification. Therefore, the United States could not officially and legally withdraw until 2020, specifically, the day after the next American election. Despite the Obama administration's strategic efforts, nothing was left to prevent American negotiators from torpedoing international meetings while still members of the Agreement, in line with the incumbent Trump administration's political agenda.

While there are legal rules and administrative obligations in the Agreement, the fact remains that without real punitive measures in place, nothing prevents a country from withdrawing or breaking its own commitments. There are no real sanctions, only moral condemnations, which have no effect on the Trump administration.

The United States' withdrawal gradually eroded the spirit of collaboration and determination to face the rising global challenge of climate change. President Trump also withdrew the United States from the WHO, UNESCO, and other international organizations without hesitation, creating a major vacuum in multilateral relations.

In my view, and admittedly with the benefit of hindsight, this withdrawal without sanctions had a historic impact on the fight against climate change. It signalled that countries that signed the Paris Agreement could break their commitments with no real consequences. That conclusion had two paradoxically opposite effects.

First, countries that signed the Agreement reluctantly, such as fossil-fuel producers and those concerned by the profound societal changes required, were empowered to challenge, delay, or even hijack negotiations during COPs. With one of the planet's biggest GHG emitters conspicuously absent, the conferences no longer displayed the same sense of unity. The fear of sanctions or diplomatic pressure from the world's leading power (and often economic partners) during negotiations disappeared, with President Trump loudly decrying the fight against climate change. He managed to get elected in a country renowned for its democratic and social values, a country that, under President Obama, had succeeded in rallying China and other emerging countries that had long been resistant to a global climate agreement. For many countries that hesitantly supported the previous Obama administration's ambitious approach, Trump sent out a strong political signal that reopened the question of commitments. Why respect ambitious GHG reduction targets, which would come with a considerable cost for local economies, when despite being the most influential player in the previous negotiations, the United States had abandoned the Agreement entirely with no sanctions whatsoever and its population had elected a climate-skeptic president?

The other political message sent by the US withdrawal without sanctions was more subtle and more pernicious. The Trump administration clearly showed that targets countries announced as part of the Agreement might not have as much moral value as expected. This conclusion opened the door to certain countries deviating from their commitments, given that one of the planet's biggest GHG contributors had abandoned its targets entirely without consequences. It eroded values and moral commitments, since the solemnity of the promises made was no longer being respected. Some countries backtracked on their implementation plans, while others made unrealistic announcements.

For example, countries looking to gain political capital by establishing themselves as leaders in the fight against climate change might announce ambitious reduction targets at major international conferences, but fail to provide the resources or implementation programs to match their stated goals. Political leaders who portrayed themselves as saviours of the planet would, for the most part, no longer be in office when the time came to be held to account, since reduction targets are medium- and long-term projections. By that point, failure to meet their ambitious international GHG reduction targets would just be one factor among many. The parade of promises from some countries clearly showed that they already knew they would not be the only ones to miss their targets and have to face public opinion. Since the Paris Agreement is not legally binding, that would be the only moral sanction to fear, and for some leaders, it was a small price to pay. They were also well aware that full compliance with the reduction timetable required by science and approved by the Paris Agreement would entail major economic compromises at the national level. What are international commitments truly worth when political leaders return home and are confronted by industrial lobbies that drive the local economy and are resistant to change? The rhetoric about the importance of bold action against climate change makes for good press, but how often is it actually realistic and accompanied by concrete measures and efforts to achieve it? In many cases, it seems to be nothing more than lip service.

Despite President Trump's decision coming as a slap in the face to multilateralism, some leaders stood up and implored countries to remain in the Paris Agreement. There was also some pushback from other levels of government. In the United States, the governors of California, Washington, and New York mobilized and announced the creation of the United States Climate Alliance, a bipartisan coalition representing almost half the American population. This group of American states intended to implement the Paris Agreement rules that fell within their jurisdiction despite opposition from the White House. As for Canada, Quebec played a key role as a leader on the international stage. Rarely had a federated state been so involved in the fight against climate change. There was considerable mobilization from subnational governments and cities across the world. Civil society was also mobilizing, and the voices of women, young people, Indigenous people, and the business world were making significant headway in UN forums. Despite this mass mobilization underscoring the urgent need to act, the fact remains that the damage to multilateralism was substantial and had an impact behind the scenes at major international forums.

In 2017, the 22nd Conference of the Parties was held in Marrakech, Morocco. COP22 was seen as the climate action conference. The aim was to start developing an implementation plan, a roadmap for achieving the many commitments of the Paris Agreement. Several issues were on the agenda and a number of questions needed to be answered to honour the commitments made in Paris the previous year. What was the best way to verify the sincerity and transparency of countries' efforts to reduce GHGs according to their NDCs, particularly after the United States announced its withdrawal? How should the commitment record be managed? What was the best way to follow up on countries' five-year plans? What was the best way to supervise developed countries' pledge to contribute $100 billion a year in international aid for developing countries starting in 2020?

Discussions progressed well, despite the American elephant in the room. Countries observed and studied one another, but none tried to hold the negotiations hostage. COP22 in Marrakech took several steps forward, marking significant progress.

The following year, demonstrations were held in the streets during COP23 in Bonn, Germany. Negotiators needed to work on proposals for texts to govern the implementation of the Paris Agreement. This was the last chance to add provisions to make countries' commitments binding. Of course, it would be exceptional for a rulebook governing an Agreement without punitive measures to overstep its mandate, but civil society had been denouncing the non-punitive nature of the Agreement, particularly since the United States' withdrawal. Tens of thousands of people demonstrated in the streets of Bonn. In their eyes, there were still no concrete results, no reduction in GHGs, despite this being the 23rd international climate conference. There were no great expectations for COP23; it was what is known as a technical COP. Progress needed to be made on the book of procedural rules, which was due to be published at the following climate conference in Katowice, Poland.

In December 2018, the 24th Climate Conference took place in the heart of Poland's coal basin. To get to the summit from the airport, participants had to pass industrial sites with tall chimneys belching black smoke, which clashed spectacularly with the theme of the conference. The Polish COP24 presidency was quickly overwhelmed by the scale of the task ahead. Multilateralism was once again facing considerable headwinds. Brazil had just elected a climate-skeptic president, Jair Messias Bolsonaro, who announced his refusal to host the next COP25 in his country less than a year before the event, which had already been planned in the meeting schedule. In France, the *gilets jaunes* movement gained traction just before the opening of COP24. The working-class movement was sparked by a rise in fuel prices due to a sudden increase in France's carbon tax. The proposed ecological transition lacked the incentives and support measures needed for the change to be accepted by the public. It was a pivotal moment that had a major international impact, sending a strong message and acting as a deterrent for many governments regarding their forthcoming climate change policies. Concepts of social justice were increasingly coming into play in drafting the rules of procedure, which were still under negotiation. With such headwinds, the spirit of collaboration that led to the ratification of the Paris Agreement and the universal appetite for ambitious climate change action were falling apart. President Bolsonaro's Brazilian delegation torpedoed the negotiations with abusive and negative obstructions.

The Intergovernmental Panel on Climate Change (IPCC), the voice of science, had just published an alarming report on the climate situation. Commissioned to guide policymakers, the scientific document recommended that countries significantly increase their GHG reduction commitments, given soaring GHG emissions and the anticipated impacts of climate change on vulnerable populations. The report estimated that if all the commitments of the Paris Agreement were respected (which was an extremely optimistic, even utopian hypothesis given the geopolitical context) the planet would be heading for warming of 2.7°C to 3.5°C, a far cry from the 2°C maximum threshold. It was vital to heed the science and act quickly to demand even more ambitious goals from countries. Unfortunately, the United States (still technically a member under the Agreement's rules of procedure), Russia, Kuwait, and Saudi Arabia joined forces during the negotiations to downplay the impact of the scientific policy paper. No additional commitments were announced, to the disappointment of civil society and ambitious countries.

The rulebook governing the Paris Agreement's implementation had been finalized, but several points remained to be negotiated. The final text of the Conference was disheartening, considering the objectives that the countries set themselves in 2015. While multilateralism survived the adverse geopolitical context once again, it failed to meet the aspirations and promised results detailed in its historic agreement.

Given Brazil's withdrawal and the unstable political situation in Chile, the next intended host of the international climate meeting, COP25 ended up being held in Madrid, Spain. The negotiations were difficult and there were points of contention that hampered decision-making, such as Article 6 of the Paris Agreement on implementing market mechanisms. Discussions on stepping up the parties' reduction targets to meet the objectives of the Paris Agreement did not lead to any commitments. However, COP25 did result in a link being established between climate and biodiversity, as well as progress on a number of important issues, such as the principle of "loss and damage," the important role of women and gender equality, and human rights.

Following that, the world was unable to gather to discuss the climate again until November 2021, due to the COVID-19 pandemic that rocked the entire planet. After two years of fruitless virtual meetings, countries and members of civil society were finally able to convene face-to-face in Glasgow, Scotland, for COP26. It was time to determine whether countries' commitments would enable them to meet the Paris Agreement's set long-term objectives. A preliminary summary report took stock of the NDCs. The analysis showed that, despite a large majority of countries (116) submitting revised NDCs, cumulative commitments were still well short of the targets required to comply with the Paris Agreement. Rather, emissions were on track to rise 13.7% by 2030 (compared to 2010), resulting in an average temperature increase of at least 2.7°C! Scientists called for a 45% cut in emissions by 2030 to maintain the Paris Agreement's 1.5°C target. The gulf between objectives and reality was widening, and the path humanity was taking would bring great risks and challenges.

With strict health regulations in place, negotiations finally resumed in Glasgow. A total of 196 countries and 120 heads of state took part in the ambitious event. COP26 President Alok Sharma insisted that countries commit to carbon neutrality by 2050. If the parties were not able to guarantee their commitments to reach "net zero emissions" by mid-century and establish reliable interim targets, it would be extremely difficult to maintain the Agreement's 1.5°C objective. Based on the preliminary analysis of the NDCs, the planet was heading for temperature increases that would have devastating effects. Given that, there was a lot of talk about adapting to climate change, mobilizing climate funds, and working with other countries to rapidly scale up climate action.

The negotiations made progress on a number of issues, and the COP presidency managed to maintain a degree of optimism among the Parties. The urgent need to act and the slow progress of ambitions seemed to be having some effect on the negotiations. In the final version of the text before its approval, the countries even managed to address the delicate issue of generating power from coal and fossil fuels. For the first time in the history of international climate agreements, the source responsible for the most GHGs was explicitly named: coal. The text also mentioned the driving force behind global warming: fossil fuels. The term had been withdrawn from the Paris Agreement during the 2015 negotiations following pressure from Saudi Arabia. This time, the negotiators succeeded in naming the main cause of climate change.

Negotiations ran over the scheduled timeframe, and there was great disappointment at the final plenary session. The final declaration document, dubbed the Glasgow Climate Pact, was amended several times to achieve the elusive consensus that is so central to proceedings.

The last-minute changes considerably weakened the commitments, with one sentence in particular taking days of intense negotiation. Negotiators had agreed to "accelerate the phasing out of coal." In the final stretch, India and China succeeded in having the words "accelerate the phasing-out" replaced by "accelerating efforts towards the phasedown," considerably weakening the scope of the commitment.

The negotiators had also succeeded in including the concept of phasing out fossil fuel subsidies. Yet again, the result was disappointing, with the final text encouraging member countries to "[accelerate] efforts towards the phasedown of unabated coal power and phase-out of inefficient fossil fuel subsidies." Not only did the text fail to address the *elimination* of fossil fuel subsidies, but the term "inefficient" was added without defining its meaning. Lastly, there was no fixed deadline or timeline for the phasedown objective. Some NGOs called it a mere public relations stunt. However, many countries considered it a significant step forward, given that the Glasgow Climate Pact, unlike the Paris Agreement, succeeded in naming the main sources of global warming.

The next three COPs were held in oil-producing countries—Sharm el-Sheikh, Egypt (2022), Dubai, United Arab Emirates (2023), and Baku, Azerbaijan (2024)—to numerous public protests. The UN was accused of caving to oil industry lobbies, which were unsurprisingly defending their interests behind the scenes, slowing down the energy transition vital to the success of the Paris Agreement.

Announcements continued to multiply with few tangible results. The harsh but inescapable reality was that since countries had made their ambitious commitments in Paris in 2015, there had been no reduction in GHG emissions. In fact, the catastrophic effects of climate change were starting to be felt across the planet, costing tens of billions of dollars. Despite that, society continued to promote economic growth at all costs, without daring to adopt a truly sustainable vision for humanity. Policy-makers' priorities remained focused on the short term, and the ambition required to meet countries' commitments was simply lacking.

At COP27, countries reaffirmed their commitment to limiting the increase in global temperature to 1.5°C, despite a trajectory putting humanity at twice that target. Global warming was already at 1.2°C, but the world refused to give up. Abandoning one of the Paris Agreement's flagship measures would have been seen as a failure of the global agreement. In front of an international audience, countries continued to forcefully proclaim the need to reduce GHG emissions, but implementation plans still lacked the wherewithal to take the action required to achieve the objectives.

After decades of talks on the issue of loss and damage, a roadmap was finally announced to establish financing mechanisms to address it. Climate change was having a major impact on the poorest communities. The new loss and damage fund provided financial support for communities to adapt to climate change and financial compensation for the most heavily impacted countries that had not significantly contributed to the environmental crisis.

While the new fund's creation and announcement was good news in theory, it also raised serious concerns from representatives of developing countries and civil society. The target of $100 billion a year by 2020 was still not met, yet another major Paris Agreement announcement that had fallen behind schedule.

The following year, COP28 was held in Dubai, one of the world's largest hydrocarbon producers. Organized by the United Arab Emirates, the international meeting had the objective of drawing up the first Global Stocktake of the commitments countries made in Paris in 2015.

The stocktake covered all the points that had been addressed in negotiations. It became a launchpad for drawing up more ambitious climate action plans that must be implemented by 2025. This stocktake confirmed that overall GHG reductions were far from the trajectory set by science.

Negotiations in the meeting rooms were challenging, but COP28 President Sultan Al Jaber spared no effort to make progress on the issues under discussion. The President of COP28 wore many hats: minister of industry and advanced technology of the United Arab Emirates, chairman of a company specializing in renewable energy, and, above all, group CEO and managing director of the national oil company ADNOC (Abu Dhabi National Oil Company). His appointment was widely criticized, as was the presence of numerous lobbyists from the oil and gas industry. However, civil society's public condemnation of the president's appointment may have spurred his drive to succeed; he proved to be an effective conciliator between the parties.

After days of hard-fought negotiations, the final declaration of COP28 included a political and economic signal that was unprecedented for an international climate agreement: The countries committed to "*transitioning* away from fossil fuels." Many considered this a major breakthrough that could mark the beginning of the end for fossil fuels. While it was still a far cry from the "*phase-out*" the NGOs wanted, fossil fuels were at least mentioned in the final text of the agreement. The Dubai agreement also called on countries to commit to tripling global renewable energy capacity and doubling energy efficiency by 2030.

COP29 in Baku, Azerbaijan, got off to a controversial start. In his opening address, Ilham Aliyev, the president of Azerbaijan and host of the Climate Summit, repeatedly championed oil and gas as a "gift from God," referring to the hydrocarbons that contributed to his country's economic prosperity. Furthermore, the concept of "transitioning away from fossil fuels" was not repeated or clarified in the final text of COP29, which many negotiators found to be a disappointing step backwards.

However, after nine years of negotiations and despite difficulty, COP29 led to the adoption of a mechanism for carbon credit (internationally transferred mitigation outcomes, ITMOs) trading between governments (Art. 6.2) and the establishment of a carbon market for project developers (Art. 6.4). The decision will require a great deal of monitoring, analysis, and rigour to prevent greenwashing. Monetizing nature in the context of carbon capture comes with risks if it becomes a quantifiable economic tool that gives countries a way to simply offset their emissions. Compensation cannot replace mitigation.

COP29 was supposed to be the COP of climate finance. The Climate Fund set at $100 billion per year expires in 2025 and must be renewed or reworked to address the urgent imperative to act. The needs of developing countries amount to over $1 trillion a year. With an adverse geopolitical context and a lack of leadership from the presidency, negotiations at the summit went around in circles. The final agreement satisfied no one. In the end, the rich countries pledged $300 billion a year to aid the global fight against climate change. For developing countries, the commitment was inadequate for the challenge at hand. They expressed deep disappointment, even describing the final text as insulting. They also denounced its lack of any mention of fossil fuels.

Facing sharp criticism and threats to derail the COP29 final declaration, the parties agreed to implement the "Baku to Belém Roadmap" to achieve financing of "at least $1.3 trillion per year by 2035." Once again, the decision was to put off deadlines and commitments while making promises for the future. At COP after COP, history repeats itself and timeframes do the same, as though the world is going the wrong way and somehow getting slower and slower each time.

Countries left Baku with a mountain of work to do before COP30, which will be held in Belém, Brazil.

2.6. The Road to Belém

The world turned upside down in January 2025. Donald Trump returned to the White House, threatening the established world order. His previous victory in 2016 took everyone by surprise, likely including the president himself and his team. Ill-equipped and poorly supported, he juggled power in his first term with a certain degree of inexperience.

For his second term, the man was ready, as was his team. Surrounded by those who support his vision, he has quickly begun to unleash isolationist measures, while threatening those who oppose them. Armed with a majority in the House and Senate, the president has free rein to flout all the established structures and rules of diplomacy. He has signed an impressive number of executive orders implementing his often harebrained and unfounded election promises. He is baldly sowing fear among long-standing allies by threatening tariffs, planning to annex territories illegally, and scrapping progressive measures and policies meant to improve inclusion for minorities. Illegal immigrants are in his sights, resulting in arrest after arrest. He is exploiting the rules of law to consolidate his power, even under false pretenses. Using the fight against fentanyl at the borders as a pretext, he has claimed that national security is in jeopardy, giving him the administrative power to economically threaten his neighbours, Mexico and Canada. He has declared a national energy emergency, giving himself the power to fast-track oil and gas projects. As he announced immediately after being sworn in for his second term, "We're going to drill, baby, drill and do all the things that we wanted to […]" Yet there is nothing to justify the energy emergency decree, since the United States produces more oil than any other country and remains the world's largest exporter of natural gas. The fossil fuel industry contributed financially to Trump's presidential campaign to the tune of over US$75 million, in return for a promise to reduce or cancel environmental studies that slow down projects and increase costs. In his speeches, the president has openly criticized the fight against climate change, which he regularly describes as a hoax. Contrary to all expectations, President Trump is drawing closer to Russia and its president, Vladimir Putin, endorsing his misleading comments about the instigation of the war in Ukraine. Brazenly flouting customary decorum, he threatened President Volodymyr Zelenskyy in the middle of a press conference in the Oval Office, a disturbing scene that will remain an infamous episode in the history of international diplomacy. Imposing tariffs on the European Union, threatening to annex Canada and make it the 51st American state, regaining control of the Panama Canal, and annexing Greenland are just a few of the items on his ever-growing list of threats and hostilities. Never before has the modern world been so transformed by a single man, who is using his position at the head of the world's greatest power to bring its full weight to bear. In just his first 30 days in office, he has transformed democracy and diplomacy around the world.

His words and justifications are frequently erroneous and misleading, but truth and facts are proving to be ineffective tools against this American president. His new allies have understood that they must fall into line or face reprisals. He has placed loyalists in charge of the FBI, CIA, and other agencies that can threaten dissenting voices with punitive measures. The country's technology oligarchs and ultra-rich have become the president's partners and contributors. Musk (X, Tesla, SpaceX), Bezos (Amazon, Blue Origin, *The Washington Post*), and Zuckerberg (Meta, Facebook, Instagram, WhatsApp) were vocal opponents of Trump during his first term, but have now rallied under the threat of the man who currently controls the House and Senate. Their respective media are highly influential sources of information for the population, but editorial responsibility for conveying truth and facts no longer seems to matter. After all, it would be tricky to censor the words of their friend the president, particularly given that lies and exaggeration are a regular feature of the White House tenant's speeches. To avoid displeasing the president, these media owners have ended fact-checking programs, relaxed online content moderation rules, abandoned DEI measures, and reoriented the editorial lines of their respective media to embrace the anti-woke counter-revolution Donald Trump announced during his campaign. Disinformation will inevitably become an additional obstacle to the environmental progress the COPs are advocating. That goal will no doubt be scrapped by many members, who are already resistant to the Paris Agreement's ambitious objectives, despite the increasingly alarming scientific forecasts and the climate catastrophes that are occurring across the planet.

On his first day in office, President Trump withdrew the United States from the Paris Agreement, reprising the step he took in 2016 that made the United States the first and only country to break the Agreement. This time, he not only attacked the treaty, but also hamstrung the agencies and civil servants responsible for climate issues. In his first days in the White House, he ordered massive layoffs, slashed the budgets of environmental agencies, cancelled climate change projects, cut subsidies to clean energy programs, openly attacked wind power projects by cancelling government permits, and even had references to climate change removed from a number of government websites. Unsurprisingly, he announced the end of US financial contributions to the UNFCCC. As if that were not enough, he cut international aid to developing countries, while discrediting the independent agency responsible for distributing tens of billions of dollars to the world's poorest, the United States Agency for International Development (USAID). Several USAID programs support climate change adaptation in the most vulnerable countries. Clearly, the United States' complete abandonment of the fight against climate change will have a catastrophic impact on the planet.

In just a few weeks, President Trump severely curtailed the US government's ability to combat climate change and play its international leadership role on the issue. By stretching the legal reach of his many executive orders, he succeeded in breaking down the walls erected by the Biden administration. Conscious of the fact that the United States' withdrawal from the Paris Agreement did not have the desired effect during his first term, the president adapted. He made his strategy clear during the election campaign. Dismantling organizations responsible for climate issues, destroying archives and reference documents, and muzzling scientists were already on the list of steps planned before he was even re-elected. This obsessive opposition to climate measures is set to have devastating and historic effects. One by one, the Biden administration's climate change programs have come under attack as the Trump administration seeks to eliminate what it calls "the Green New Scam." Trump is bent on trampling American checks and balances in what increasingly seems to be a quest for revenge. He has launched a full-blown witch hunt against scientists who do not follow his ideology, trampling on the academic freedom of not only American researchers but also scientists of other nationalities who participate in international studies on the environment, public health, or equity. The *New York Times* has published a list of words that are now banned, including "climate crisis," "climate justice," "vulnerable population," "LGBTQ," and "Gulf of Mexico." Targeting scientists will have an impact on the entire population, and the United States withdrawing from major international organizations such as the UNFCCC, the WHO, and others will affect multilateralism in a direct and fully intentional way.

The United States withdrawing from the Paris Agreement sends a strong message to countries that are reluctant to implement action plans to match their commitments: Failing to comply with international treaties has no consequences. Like in his first term, President Trump is making a mockery of the moral obligations of a country participating in an international agreement.

This is the historic context in which the world will gather in Belém, in November 2025, for the COP30, the major annual climate meeting that will also mark the 10th anniversary of the Paris Agreement. Despite the headwinds that continue to threaten and potentially weaken multilateralism, the Brazilian presidency intends to urge participants to come up with a clear and ambitious response to the United States' disengagement, and most importantly, to the common challenge facing the future of humanity. The year 2024 was the hottest since records began, and it marks the first time warming exceeded the critical 1.5°C threshold. Although it will take several years above the threshold to confirm that the goal of limiting warming to 1.5°C has failed, nothing currently indicates that the global climate situation will improve. Lowering or stabilizing temperatures in the short term is simply unattainable in the current societal context. In the abstract, we can still hope that our societies will transform their relationship with the planet overnight. We can dream of truly revolutionizing our economies. But is such optimism realistic in these times of great uncertainty? Humanity is still unable to reduce its emissions, and the natural inertia of the global climate system will take at least two decades to respond positively to any progress in reducing GHG emissions. The countries' negotiators know that they are working for future generations, and that their commitments will have no immediate impact on the climate. This is not to diminish their contributions in any way, but the scientific evidence clearly demonstrates the complexity of the task ahead to ensure an acceptable legacy for our children and grandchildren.

There are plenty of other obstacles that loom for COP30, an anniversary that requires countries to present new and improved GHG reduction targets under the terms of the Paris Agreement (a recurring obligation every five years). In such a complex multilateral framework, the Brazilian presidency will have quite a challenge to tackle. The geopolitical situation and wars, the sweeping threats from President Trump and his administration, the financial precariousness of the poorest, and the rise of the right wing are all factors that could undercut countries' ambitions in the fight against climate change.

Furthermore, the United Nations is facing significant reform due to a major liquidity crisis. In an official statement a few months after President Trump's return and the United States' withdrawal from the Paris Agreement and other UN organizations, Secretary-General of the United Nations Antonio Guterres launched a vast operation, the UN80 Initiative, to refocus the institution's priorities for its 80th anniversary. Will this in-depth reform succeed in improving multilateralism, which seems to be under threat again despite its history of resilience? Time will tell.

Admittedly, the negotiations conducted under the aegis of the United Nations have not produced convincing results in recent years. They are making little to no progress on climate, biodiversity, plastic pollution, and desertification—all issues that fall within the UN's remit—despite the extremely urgent need to act. Behind the scenes at the COPs, some negotiators believe that the current system is fundamentally broken, being held hostage by a very small number of countries who oppose change and are slowing down or blocking negotiations under the rules of decision-making by consensus. Others think that climate congregations should be created to gather the most ambitious countries, while opponents argue that current multilateralism gives the poorest countries, who are the first victims of the climate crisis, a seat at the table and the ability to influence decisions along with the highest income and powerful nations. Committee after committee is formed, but tangible results remain elusive.

COP30 will be a historic moment, held in Belém at the gateway to the Amazon. It will either mark the resurgence of hopes forgotten in the wake of recent failures, or plunge multilateralism even further into chaos. In March 2025, COP30 President-Designate André Aranha Corrêa do Lago addressed a poignant but hopeful letter to the negotiators and non-governmental representatives taking part in the event. He emphasized that change is inevitable and will come by choice or by catastrophe.

In his long missive, he referred to the human values that must underpin our forthcoming negotiations, including hope and renewal, unity, resilience, kindness, and generosity. He took inspiration from the heritage of Brazil's Indigenous Peoples, specifically the concept of *mutirão*, which refers to a community coming together to work on a shared task, whether harvesting, building, or supporting one another. In sharing this precious ancestral wisdom and social attitude, the future COP30 presidency invited the international community to join Brazil in a global *mutirão* against climate change.

That is my sincere and profound wish for all of us. I will not reiterate the urgent need to act, nor will I implore governments once again to take action. In doing so, I would simply be repeating what the voice of science has been saying for over 30 years. The efforts of scientists and a large part of civil society have not yet managed to break through the economic and political strongholds of successive national leaderships.

Our imperfect but resilient multilateral trading model has not changed in the past 30 years, while the world itself is changing at a breathtaking pace. Perhaps it is time to adapt the model to the unprecedented reality of this new world, which is just as imperfect but will hopefully be resilient enough to regain its senses and reason in time to humbly claim its own survival.

Federative Republic of Brazil
COP30 President-Designate

March 10th, 2025

Dear friends,

As we move towards the second quarter of the 21st century, the international community is bound to reflect on the shared human values that hold us together: peace and prosperity, hope and renewal, consideration and gratitude, unity and connection, resilience and optimism, generosity and kindness, diversity and inclusion. These values highlight our collective spirit in a century that will test our species' ability to adapt and innovate in building a common future.

Brazil will host and preside over the 30th session of the Conference of the Parties (COP30) to the United Nations Framework Convention on Climate Change (UNFCCC) in November 2025 against the backdrop of several landmarks: COP30 will mark 20 years since the entry into force of the Kyoto Protocol and 10 years of adoption of the Paris Agreement. Much has been learned throughout the three decades of our multilateral regime. Through achievements and shortcomings, the UNFCCC has provided a mirror of humanity's greatest qualities and limitations. It has shown us how our societies, economies, and politics should work – and how they do in practice.

I am greatly honored to have been nominated by President Luiz Inácio Lula da Silva as the COP30 President-Designate. As a long-time climate negotiator, I humbly take this immense responsibility and am determined to serve the process towards COP30 and beyond in line with our shared human values and with the mission to consolidate our common legacy, whilst innovating our response to the extent needed by the climate crisis.

Cooperation among peoples for the progress of humanity

In 1988, we, the United Nations, first identified climate change as a "common concern for humankind" and decided to create the Intergovernmental Panel on Climate Change (IPCC). Our leaders listened to scientific alerts and came together in Rio de Janeiro four years later around the ultimate objective of preventing dangerous anthropogenic interference with the climate system. At the 1992 United Nations Conference on Environment and Development, the Rio 'Earth Summit,' world leaders signed the UNFCCC, defining principles and the five building-blocks for the multilateral response to climate change: mitigation, adaptation, finance, technology, and capacity-building.

Similarly to the UNFCCC's role in inaugurating multilateral climate governance, Brazil's Federal Constitution adopted in that same year of 1988 established the fundamental objectives of the Brazilian Republic: to build a free, just, and caring society; to guarantee national development; to eradicate poverty and marginalization and reduce social and regional inequalities; and to promote the good of all, without prejudice to origin, race, sex, color, age, or any other form of discrimination. Brazil's Constitution equally binds the country to be governed in its international relations by principles that include "cooperation among peoples for the progress of humanity." This fundamental principle will guide the incoming presidency of COP30 – not only because the Brazilian diplomacy is constitutionally bound by it, but because it has the firm conviction that there is no future progress for humanity without deep, rapid, and sustained cooperation among our peoples.

COP30 at the epicenter of the climate crisis

We now enter 2025 with the confirmation that 2024 was the warmest year on record globally, and the first calendar year that the average global temperature exceeded 1.5°C above its pre-industrial level. January 2025 further marked the warmest month on record. Building on previous work on physical, transitional, and legal climate-related risks, the Financial Stability Board – the international body that monitors and recommends policies for the global financial system – reported last January that climate shocks can threaten the world's financial stability. COP30 will therefore be the first to undeniably take place at the epicenter of the climate crisis, and the first to be hosted in the Amazon, one of the world's most vital ecosystems, now at risk of reaching an irreversible tipping point, according to scientists.

We have long known the scale and gravity of climate change and its growing impacts. We have affirmed and reaffirmed global warming as an existential threat to humankind. We have had scientific knowledge on the issue for over 35 years, consolidated since the first 1990 IPCC assessment report.

Now, not only do we hear about climate risks, but we also live the climate urgency. Climate change is no longer contained in science and international law. It has arrived at our doorsteps, reaching our ecosystems, cities, and daily lives. From Siberia to the Amazon, from Porto Alegre to Los Angeles, it now affects our families, health, cost of living, and routines in education, work, and entertainment. Images of climate disasters and human suffering invade our living rooms on TV and on social media, as we rapidly enter a dangerous zone in which the rich in developed and developing countries isolate themselves behind climate-resilient walls. Meanwhile, the poor in both developing and developed countries suffer more and more. Inevitably, extreme weather events – and potential climate tipping points – will increasingly affect every country, community, and individual, though the most vulnerable will be the most affected.

A global call against climate change

While we grieve human and material losses, 2025 must be the year we channel our sadness and indignation towards constructive collective action. Change is inevitable – either by choice or by catastrophe. If global warming is left unchecked, change will be imposed on us as it disrupts our societies, economies, and families. If instead we choose to organize ourselves in collective action, we have the possibility of rewriting a different future. Changing by choice gives us the chance for a future that is not dictated by climate tragedy, but rather by resilience and agency towards a vision we design ourselves.

In coming to terms with reality when countering doom, cynicism, and denial, COP30 must be the moment of hope and possibilities through action – never paralysis and fragmentation. We must face climate change together and reactivate our collective and individual ability to respond: our "response-abilities."

The Brazilian culture inherited from Brazilian native indigenous peoples the concept of "mutirão" ("Motirõ" in Tupi-Guarani language). It refers to a community coming together to work on a shared task, whether harvesting, building, or supporting one another. By sharing this invaluable ancestral wisdom and social technology, the incoming COP30 presidency invites the international community to join Brazil in a global "mutirão" against climate change, a global effort of cooperation among peoples for the progress of humanity.

Together, we can make COP30 the kickstart of a new decade of inflection in the global climate fight. As the nation of football, Brazil believes we can win by "virada." This means fighting back to turn the game around when defeat seems almost certain. Together, we can make COP30 the moment we turn the game around, when we put into practice our political achievements and our collective knowledge to change the course of the next decade. COP30 can be the COP we align efforts worldwide: from national to local governments, from international capital markets to local bazars, from major technology actors to local innovators, from academic to traditional knowledges.

Summoning the United Nations in a new alliance against our common enemy: climate change

In recovering our abilities to respond, we must tap into the inspiration of historical victories in overcoming past existential threats. 2025 is also the year the international community remembers it represents the legacy of the alliance that eight decades ago chose to leave differences behind to unite against the scourge of war. This year marks the 80th anniversary of the end of World War II and of our alliance in creating the United Nations. German-American philosopher Hannah Arendt denounced the "banality of evil" as the acceptance of what was unacceptable. Now, we face the "banality of inaction," an irresponsible and unacceptable inaction.

In this critical decade, Brazil summons back our alliance of peoples to once again leave our differences behind and unite in vanquishing our common enemy: climate change. This time, we will count on strong foundations to lead us to victory. Science confirms we have the resources to combat climate change. Among them, our technology now taps into life and digital networks that can connect, leverage, and distribute resources through unprecedented flows in speed and scale. Though greatly inequitable and vulnerable to climate risks, our financial architecture boasts sophistication gained from previous crises that can be reformed and improved further. COP30 can be the moment we align international financial flows and merge the digital and climate transitions into one single new industrial revolution that is climate conscious.

Pulling levers: calling brilliant minds, brave individuals, hard work solidarity

To leaders and stakeholders beyond the UNFCCC – in finance, subnational governments, private sector, civil society, academia, and technology – the incoming COP30 presidency invites you to join our global "mutirão." Humanity needs you.

3

Ancient Greek mathematician and physicist Archimedes said, "give me a lever long enough and a fulcrum on which to place it, and I shall move the world." To leaders and stakeholders in all walks of life, give us levers long enough and COP30 will serve as a fulcrum on which to place them. Together, we shall move the world towards low-carbon and climate-resilient transitions.

To our thinkers, spiritual leaders, artists, and philosophers, we call on you to help us transcend outdated mindsets whilst preserving shared values and innovating towards a new planetary renaissance. Humankind must regenerate its relationship with itself and with the nature it belongs to.

To local leaders, small businesses, parents, individuals, and professionals in health, education, and public safety, we need you to regenerate our communities as strongholds of belonging, cooperation, and purpose. Our human family will only be as resilient as our communities and neighborhoods are cohesive and strong. In enhancing the values of citizenship, we need to offer our children vision, exemplary models, and mentorship in demonstrating that the extent to which we respect each other, and our environment, is the extent to which we respect ourselves.

As we prepare for COP30, the incoming presidency will be recruiting actors among non-State stakeholders to partner-up as "levers" in helping apply solutions to "high leverage points," where small changes can result in large impacts on complex systems' behaviors. Incentives representing the rules and boundaries of our systems can become strong leverage points for just, fast, and comprehensive climate transitions.

The recognition of the need to act as soon as possible to face the urgency of climate change should inspire new attitudes. We must recognize that issues considered "problems" can emerge as important "solutions." We can revert the perception of the role of some actors, sectors, technologies, and practices that have evolved and, by being already available, can represent important contributions.

When we get together in the Brazilian Amazon in November, we must listen to the latest science and re-evaluate the extraordinary role already played by forests and the people who preserve and rely on them. Forests can buy us time in climate action in our rapidly closing window of opportunity. If we reverse deforestation and recover what has been lost, we can unlock massive removals of greenhouse gases from the atmosphere while bringing ecosystems back to life. Healthier ecosystems can equally offer resilience and bioeconomy opportunities by promoting local livelihoods, creating sophisticated value-chains, and generating innovations in biotechnology. Tapping into such an outstanding potential requires enhanced global support and investment, including through financial resources, technology transfer, and capacity-building.

Pulling levers: calling stakeholders within the UNFCCC

To leaders and stakeholders within the UNFCCC negotiations, COP30 must represent a decisive transition from the regime's negotiation phase, which has already succeeded in positioning climate at the center of the world's economic, social, and political debates. Significant collective progress towards the Paris Agreement temperature goal has been made, from an expected global temperature increase over 4 °C, according to some projections prior to the adoption of the Agreement, to an increase in the range of 2.1–2.8

°C with the full implementation of the current nationally determined contributions (NDCs).

The Paris Agreement is working, but there is much more to do.

To climate negotiators, as we continue to reinforce the regime, it is relevant to be self-critical and act upon much of the outside perception of talks having lingered for over three decades with meager results. In view of climate urgency, we need a new era beyond negotiating talks: we must help put into practice what we have agreed. We must decisively pull the levers of our processes, mechanisms, and bodies towards aligning efforts within and outside the UNFCCC with the long-term goals of the Paris Agreement on temperature, resilience, and financial flows.

To national policymakers and political leaders, governments have the response ability to pull the levers of climate action and ambition in their next NDCs. In integration, our NDCs must align with the temperature goals of the Paris Agreement. National leaders must honor their resolve to pursue efforts to limit the temperature increase to 1.5 °C. Human lives depend on it, future jobs depend on it, healthy environments depend on it.

There is a high expectation of taking stock of NDCs at COP30. As we all know, NDCs are nationally determined and hence not subject to multilateral negotiations. We will nevertheless stimulate a frank collective reflection on bottlenecks that have been hampering climate ambition and implementation.

We will be judged in the future by our willingness to firmly respond to the growing climate crisis. Lack of ambition will be judged as lack of leadership as there will be no global leadership in the 21st century that is not defined by climate leadership. We can be on the right side of history by turning NDCs into platforms for a prosperous future that enshrine national determination to contribute and transform. In the run-up to COP30, we need ambitious NDCs that privilege quality as a follow-up to legal obligations under the Paris Agreement.

In helping each other in transitions that are just, our common but differentiated responsibilities will serve as strong levers for countries' willingness to contribute to the climate fight. Our fulcrum: international cooperation for strengthening respective capabilities and institutions in all countries. As we acknowledge we are all interdependent in the fight against climate change, we must recognize the international community is only as strong as its weakest link.

COP30 will serve as a fulcrum for long levers within and beyond the UNFCCC because our multilateral climate regime is strong, resilient, and resourceful. Climate multilateralism boasts the wisdom and achievements from each of the past twenty-nine COPs. Standing on the shoulders of our predecessors, the incoming presidency of COP30 is humbled by the legacies of COP21 to COP29, legacies that we must preserve and build upon.

Upholding multilateralism: preserving and expanding our collective legacy

Supported by the entire UN System, as determined by Secretary-General António Guterres, our multilateral institutions can and will deliver results commensurate with the scale of the climate challenge.

Ever since Brazil received the trust of Latin America and the Caribbean as our region's COP host, the path to COP30 has been successfully paved by the COP28 and COP29 Emirati and Azerbaijani presidencies – our troika partners in the Road Map to Mission 1.5. In 2023, under Emirati leadership in Dubai, we adopted the UAE Consensus, which included a breakthrough on loss and damage, following the Egyptian COP27 leadership, and the conclusion of the first global stocktake (GST). Unprecedentedly, the GST launched global calls for efforts towards halting and reversing deforestation and forest degradation by 2030, and for accelerating the global energy transition, including by tripling renewable energy capacity globally, doubling the global average annual rate of energy efficiency improvements, and transitioning away from fossil fuels in energy systems, in a just, orderly, and equitable manner.

Based on equity and science, the GST is already the unanimous reference that informs international cooperation and Parties in enhancing actions and support. The GST stands as our guide to Mission 1.5, as our collective project to implement the vision of the Convention and the Paris Agreement, the vision of strengthening the global response to the threat of climate change, in the context of sustainable development and efforts to eradicate poverty.

Having COP28 as a steppingstone and then under the Azerbaijani leadership of COP29, we finally completed in 2024 the Paris "Rulebook" by finalizing the rules under Article 6. We further adopted the "Baku Climate Unity Pact," which includes the cornerstone decision around the new collective quantified goal on climate finance (NCQG). The incoming COP30 presidency looks forward to working with the COP29 presidency in guiding the "Baku to Belém Roadmap to 1.3T" to scale up climate finance to developing country Parties. Together, we will be producing a report summarizing our work by COP30. The "Baku to Belém Roadmap to 1.3T" must serve as a fulcrum for leveraging finance to low-carbon and climate-resilience pathways in developing countries, recalling that IPCC's alerts on the urgency of climate action are centered on findings that finance, technology, and international cooperation are critical enablers for accelerated climate action. Experts are clear: we only have a few years. If climate goals are to be achieved, both adaptation and mitigation financing will need to be increased manyfold.

Climate change represents one of the greatest challenges of our time and addressing it should be spearheaded by progress towards sustainable development and the mobilization of all of humanity's resources to tackle structural inequalities within and among countries, while paving the way for just transitions towards low-carbon and climate-resilient societies. Though this may sound idealistic, the reality is that there is sufficient global capital to close the global investment gap but there are barriers to redirecting capital to climate action. Governments, through public funding and clear signals to investors, are key in reducing these barriers. We need to use in the best way the multilateral financial architecture and remove barriers and address disenablers faced by developing country Parties in financing climate action, including high costs of capital, limited fiscal space, unsustainable debt levels, high transaction costs, and conditionalities for accessing climate finance. We must progress in mainstreaming climate into investments and finance.

In guiding the "Baku to Belém Roadmap to 1.3T" alongside the COP29 presidency and in consultation with Parties, the incoming COP30 presidency reiterates the call on all actors to work together to enable the scaling up of financing to developing country Parties for climate action from all public and private sources to at least USD 1.3 trillion per year by 2035. It is high time Multilateral Development Banks (MDBs) and International Financial Institutions (IFIs) evolved into bigger, better, and more effective entities that structurally support enhanced, ambitious climate action.

Navigating ahead: guided by the Southern Cross

As we step into 2025, we go from COP29 to COP30 not only with a complete Rulebook for the Paris Agreement but also with its policy cycle fully in motion, including in terms of NDCs and the enhanced transparency framework (ETF). We do have pending issues to solve at COP30, notably the UAE dialogue on implementing the GST outcomes and the just transition work programme (JTWP). The GST is an invaluable legacy that unites us. We must all continue to subscribe to it as the ultimate benchmark for climate implementation. Just transitions are central for leveraging climate action towards sustainable development and addressing structural inequalities between and within countries, including in terms of gender, race, and ethnicity.

In previous COPs in the Northern Hemisphere, we navigated guided by the "North Star." As COP30 moves to the Southern Hemisphere, we look at the sky to find the five stars of the "Southern Cross" as our compass in reaching decisive inflexions across all the UNFCCC's five pillars – mitigation, adaptation, finance, technology, and capacity-building. Parties have recognized in Baku we must double-down efforts to support just transitions across all sectors and thematic areas, and cross-cutting efforts, including transparency, readiness, capacity-building, and technology development and transfer. As countries prepare and communicate their next NDCs and Biennial Transparency Reports (BTRs), we need to build the capacity of developing country Parties to transition from ad hoc reporting approaches to government-led, systematic, and institutionalized processes for preparing and submitting national reports under the ETF. The mandated workshop to be held at the sixty-second session of the Subsidiary Body for Implementation (SBI62, June 2025) will facilitate the sharing of experiences of developing country Parties in preparing their first BTRs. SBI62 will be equally key to the elaboration of the technology implementation programme and the review of the Climate Technology Centre. Parties are expected to agree on the technology implementation programme at COP30 to strengthen the UNFCCC Technology Mechanism and support the implementation of technology priorities identified by developing countries.

In 2025, we will continue and strengthen the Sharm el-Sheikh dialogue on the scope of Article 2, paragraph 1(c) of the Paris Agreement. The Paris Committee on Capacity-building (PCCB) will develop a work plan, whilst its focus for the year relies on capacity building for designing holistic investment strategies, bankable projects, and stakeholder engagement to strengthen the implementation of NDCs and National Adaptation Plans (NAPs) in developing countries. SBI62 will also initiate the development of a new gender action plan, taking into account the review of the enhanced Lima work programme. Brazil is honored to build on the legacies of previous Latin American COP presidencies with a view to advancing the agenda on gender and climate at COP30.

At COP30 we also have the unique opportunity of amplifying the invaluable legacies from the British COP26 and the Egyptian COP27 leaderships, including under the Sharm el-Sheikh mitigation ambition and implementation work programme (MWP). Instead of suspicion in polarized negotiations, the MWP has the vocation of becoming a platform for breakthroughs and trust-building through action and cooperation when leveraging opportunities, overcoming barriers, and exploring actionable solutions.

Let us bring the "mutirão" spirit to the mitigation ambition and implementation work programme. Discussions were held in Baku around the creation of a digital platform to facilitate implementation of mitigation actions by enhancing collaboration between governments, financiers, and other stakeholders on developing investable projects in a country-owned and nationally determined manner. Such a digital platform can serve as a fulcrum for powerful levers in climate implementation, with speed and scale.

As we face and recover from extreme climate events around the world, we must ensure 2025 is equally a landmark for climate adaptation and the delivery of NAPs. Governments, businesses, subnational stakeholders, financial institutions, and universities need to put adaptation at the same level of engagement and centrality as mitigation. Adaptation is no longer a choice, nor does it compete with mitigation. Building on progress on the global goal on adaptation (GGA) at COP28 and COP29, we must fulfill our legal mandate on indicators under the United Arab Emirates–Belém work programme.

Advancing the "Baku Adaptation Road Map" and the Baku high-level dialogue on adaptation will be essential, side by side progress on the work on the Santiago network and the Warsaw International Mechanism. Climate realism requires adaptation to be at the forefront and center of everything we do as governments, private sector, members of civil society, and individuals. A major inflexion on adaptation at COP30 will be the gateway for aligning our multilateral process with people's daily realities: climate adaptation is the vehicle for care and repair towards collective transformation.

The more ubiquitous our fight against climate change becomes, the more we need to incorporate synergies between climate, biodiversity, desertification, and our Sustainable Development Goals (SDGs). The 1992 Earth Summit was the cradle of the Rio Conventions, the Rio Declaration and Agenda 21. Twenty years later, our leaders gathered again in Rio, at the 2012 United Nations Conference on Sustainable Development (Rio+20), around "The Future We Want," which culminated in the SDGs in 2015, in the same year we adopted the Paris Agreement.

Back to Brazil – and now in the Amazon – these agendas need to be integrated through strong public participation. It is urgent we address, in a comprehensive and synergetic manner, the interlinked global crises of climate change and biodiversity loss in the broader context of achieving the SDGs. In doing so, we must continue to acknowledge and expand the role and contributions of Indigenous Peoples and of local communities in nature stewardship and climate leadership, while recognizing the disproportionate effects they suffer from climate change.

A new era: honoring our word

As negotiations emanating from COP21 conclude, we must refocus our efforts on action and implementation. Words and text must be translated into actual practice and transformations on the ground. The credibility and strength of the regime hinge upon it. COP30 must mark the moment we transition to the UNFCCC "post-negotiation" phase. We must intensify the consideration of approaches and initiatives to "increase the efficiency of the UNFCCC process towards enhancing ambition and implementation," including through related ongoing work under the SBI.

With climate urgency, the complexity of our task ahead is to strengthen climate governance and provide agility, preparedness, and anticipation in both decision-making and implementation. For channeling collective wisdom, the incoming COP30 presidency will invite all presidencies from COP21 to COP29 to form a "Circle of Presidencies" for advice on the political process and on climate implementation. We will further invite the current presidencies of COPs under the UN Convention on Biological Diversity (CBD) and UN Convention to Combat Desertification (UNCCD). The "Circle of Presidencies" will help ensure COP30 honors and synthesizes the legacies of previous COPs while reflecting on the ongoing agenda and the future of our process and of global climate governance. In synergy with the sustainable development, biodiversity, and desertification global agendas, the circle can leverage networks and articulate resources, processes, mechanisms, and stakeholders within and outside the UNFCCC, to make the difference locally when aligning with the Convention and the Paris Agreement. We will also invite leaders among Indigenous Peoples to form a "Circle of Indigenous Leadership" to help integrate traditional knowledges and wisdom into global collective intelligence.

The incoming presidency will continue to engage and count on the Mission 1.5 Troika to significantly enhance international cooperation and the international enabling environment to stimulate ambition, action, and implementation over this critical decade and keep 1.5 °C within reach. The COP30 presidency will work closely with the UN Secretary-General on raising climate awareness and political momentum around NDCs, on promoting information integrity about climate change, and on fostering public mobilization, including in the context of the partnership between Secretary-General Guterres and President Lula.

The incoming presidency will also undertake a "Global Ethical Stocktake" (GES) to hear from a geographically diverse group of thinkers, scientists, politicians, religious leaders, artists, philosophers, and traditional peoples and communities, among others, about ethical commitments and practices for dealing with climate change at all levels. As French philosopher Rabelais warned us in the 16th century, *"science sans conscience n'est que ruine de l'âme"* ("science without conscience is but the ruin of the soul").

Finally, the incoming presidency will engage in the months ahead on a series of further collective initiatives eyeing long-lasting positive impacts. In addition to desired progress on areas already mentioned, we will be making other announcements in the coming weeks and months, seeking to bring to the Action Agenda a new dynamic that focuses on key issues for the full implementation of the GST and of NDCs.

As the COP comes to the Amazon, forests will naturally be a central topic. Building on the outcomes of the 2023 Amazon Summit, the initiative "United for our Forests" will stimulate the debate on the role of forests in the climate fight. Other themes to receive special attention will include energy, cities, and technology and innovation.

The incoming presidency will bring together negotiators, governments, civil society, private sector, and other stakeholders to engage in an exercise on how to translate the 10 years of Paris into palpable achievements and incentives to continue to act and strengthen the multilateral regime. A group of Special Envoys will similarly engage with key actors to integrate different solutions and dimensions of the climate challenge which remain fragmentally addressed. In 2026, the Brazilian presidency will follow-up on these efforts in coordination with the future incoming COP31 presidency.

Beyond marking an isolated event, COP30 must respond to the climate crisis by triggering a "movement of movements" – a global movement of local, multistakeholder, and multisectoral movements. The integration of movements into a global movement will aim at incorporating to the preparations of COP30 principles of complexity science: <u>together, we can make the whole of our efforts "emerge" as more than a mere sum of their parts</u>. More importantly, such a global movement will be able to recover our <u>sense of shared destiny</u>.

Building tomorrow, making history today

COP30 will mark the midway in humanity's critical decade in the fight against climate change as our common enemy. Now is the time we leave behind inertia, individualism, and irresponsibility to embrace the best versions of ourselves through creativity, solidarity, and perseverance. Countries, businesses, and individuals that anticipate the radical changes ahead will be those who prosper by building resilience and tapping into opportunities in engaging, innovating, and adapting.

We live in a historical moment. Systemic climate-related risks are already progressively showing signs. Climate shocks may not come slowly – they might emerge abruptly, in irreversible shifts.

In our fight against climate change – the fight of the century – all actors and every product and service will come under scrutiny everywhere, now and in the future. Those who refuse to reflect mid- to long-term thinking, future-oriented policies and engagement may succumb to reputational, transitional, legal, and physical climate-related risks. Those who genuinely commit to winning the climate fight have the potential to emerge in leadership in a golden age of global renewal, regeneration, and cooperation.

The incoming COP30 presidency is determined to serve as a platform for collective organization and mobilization, a vessel in a global "mutirão" against climate change. Let us pull the levers together. Let us move the world.

André Aranha Corrêa do Lago
COP30 President-Designate

Figure 2.1. Letter from the Brazilian presidency of COP30.

CHAPTER 3

So COP, What Have You Done with Your Life?

Richard Kinley

Many of us, as we approached the milestone of our 30th birthday, used the occasion to take stock of what we have achieved in life and to look ahead to what needs to be done, or adjusted, to realize our goals. And so it should be with the United Nations Framework Convention on Climate Change (UNFCCC) Conference of the Parties (COP) as the 30th such event looms on the horizon. This is even more the case given the reality that, despite almost 30 COPs, the problem they were established to deal with and to help solve—climate change—has accelerated to become a multifaceted global emergency.

In my almost 25 years as a senior United Nations official, I had the privilege of helping governments to negotiate on, and find consensus in, achieving their stated objective of coming to grips with the climate challenge. This involved organizing COPs in ways that promoted successful outcomes and advising COP presidents on how to organize and manage conferences and their constituent negotiating processes and events. It was a career marked by precious moments of jubilation and satisfaction, long periods of frustration and despair, and occasional moments of historic importance.

In this chapter, I present a view of the evolving character of COPs, including a chronology of negotiations, assess the situation heading into COP30, and identify options for how COPs could make more effective contributions to resolving the climate crisis.

3.1. Let's Take a Step Back

Before proceeding, it is useful to examine the context in which COPs occur, especially as they are only part of a wider "regime," namely the collection of international institutions and processes that governments are using to help address the problem of climate change. This regime grew out of the first comprehensive scientific assessment of climate change presented by the Intergovernmental Panel on Climate Change (IPCC) and the ensuing 1990 decision of the United Nations General Assembly (UNGA) to launch negotiations on a climate change treaty.[1] The result of the ensuing negotiations, the 1992 Framework Convention on Climate Change, has led to two further agreements under the Convention—the Kyoto Protocol (1997) and the Paris Agreement (2015). The UNFCCC process sits at the centre of the global climate regime, and encompasses the COPs, their subsidiary bodies, numerous constituted bodies and processes (e.g., reporting and review), and a financial mechanism. It has become a sprawling and complex process involving almost continuous meetings and activities that engage tens of thousands of experts, negotiators, and advocates throughout the year.

---

1. The IPCC was established by the United Nations Environment Programme and the World Meteorological Organisation in 1988 to provide scientific assessments of climate change. Its first assessment report (1990) provided the scientific foundation for the emerging global climate negotiations that were launched by UNGA resolution 45/212 on the "protection of global climate for present and future generations of mankind" (December 1990). It is notable that science was a foundation for launching the negotiations and that the scientific knowledge (of climate change, and scientific consensus on the nature and causes of climate change and its impacts as well as actions possible or required) has grown exponentially through six assessment reports.

Moreover, the UNFCCC process is not the sole international actor in this regime, as climate action requires the engagement of many international organizations and processes within and outside the United Nations,[2] as well as a multitude of intergovernmental and other activities and events from the UN and the UNGA to the multilateral development banks all the way to small regional and international sectoral organizations. This complexity of the regime befits an issue as socio-economically complex as global climate change, for which there is no simple solution.

3.2. What Is the COP Really?

The character and purpose of the COP are several and have evolved over time. I will focus on five such purposes, while acknowledging that there may be even more.

a) The COP is a membership meeting

The Conference of the Parties is first a meeting of the "members" of the "climate club," namely the states that have ratified the Convention, the Kyoto Protocol, and the Paris Agreement (aka the "parties").[3] They meet annually, unless the COP decides otherwise.[4] Almost all states are "party to" all three of the climate treaties; this universality is a huge achievement and indicates a first level of commitment to climate action. At the same time, the range of national interests and

---

2. Examples include the FAO, ICAO, and IMO within the UN system while outside one could mention OECD, IEA, G7, G20, and BRICS.

3. There are in fact three "COPs." Because the membership of the three treaties varies slightly, each treaty requires its own decision-making body or COP (abbreviations: COP, CMP [Kyoto], CMA [Paris]). In the Secretariat, we made considerable efforts to integrate the three COPs to the extent possible in order to maximize efficiency and coherence within the legal constraints.

4. Which the COP has never done; only in 2020 was the COP postponed due to the COVID-19 pandemic.

priorities within this membership is breathtaking. Anyone who has struggled to bring a small meeting to a consensus will sympathize with a COP president who has to keep over 190 members on board to adopt decisions by consensus (as voting rules have never been adopted).

There have always been observers in attendance at COPs. These representatives of civil society, business, and other non-state actors are not full members with decision-making rights but have become increasingly active participants in COPs. The scale of their participation in the process has grown steadily since COP1 (Berlin, 1995), with a virtual explosion in numbers in the last 10 years. This is one factor, along with the heightened priority attached to the climate change issue since 1995, that has contributed to the evolution in the character of COPs away from rather straightforward annual meetings for the purpose of diplomatic negotiations among states.

b) The COP is a negotiating forum

Negotiation has traditionally been the most important function of COPs: negotiation of new agreements (à la Kyoto and Paris), of the rules guiding implementation of the agreements and the wider climate change regime, and of the COP decisions signalling directions or actions. Agreements are treaties and are therefore legally binding on the parties having ratified them; COP decisions, on the other hand, are not legally binding but have high levels of authority, especially when they use language of compulsion.[5] The negotiation of two significant and evolutionary universal agreements (in addition to the Convention), along with thousands of pages of decisions, shows that the COP is a powerful negotiating forum. Further details on the negotiations are provided in Section 3.3 below.

---

5.   One of the characteristics of the climate regime, however, is its proclivity to employ verbs of encouragement rather than compulsion in agreements and in decisions. For example, "should" or "may" rather than "shall"; "urges"; "invites"; "with the aim of," etc.

Canada's Role – Process over Action

In the late 1980s, Canada played a leading role in bringing the climate change issue into public focus. The Toronto Conference on the Changing Atmosphere (June 1988) was a very important step in the process of building momentum. In his speech to the conference, Prime Minister Brian Mulroney called for the negotiation, by 1992, of a framework convention for the protection of the atmosphere. Also notable was the February 1989 Ottawa meeting of legal and policy experts on protection of the atmosphere.

Canada's record in the UNFCCC process, on the other hand, can generously be characterized as "mixed." On matters relating to action or ambition Canada has been seriously constrained by its emissions realities and domestic political challenges/failures. This was most vividly demonstrated by its embarrassing need to legally withdraw from the Kyoto Protocol in 2011 due to failure in compliance. Where Canada has been able to show leadership is in process participation. It successfully hosted COP11 (Montréal, 2005) with Stéphane Dion, then minister of environment, serving as resident. In a similar vein, many Canadian delegates have served as presiding officers of thematic UNFCCC consultation processes and helped to broker understandings big and small. An excellent example from the early years is Elizabeth Dowdeswell (then an Environment Canada assistant deputy minister and later executive director of UNEP) who served as co-chair of one of the working groups negotiating the Convention (1991-1992). Similarly, then Minister of Environment and Climate Change Catherine McKenna co-led ministerial consultations for the COP presidency leading up to and at COP21 where the Paris Agreement was agreed. She also famously, and somewhat misleadingly in terms of action, declared that "Canada is back" on the eve of Paris.

c) The COP is an awareness-raising moment

Each annual COP is an opportunity to raise awareness about the climate change issue, about the range of solutions to the problem, and about the need to adapt to the impacts of climate change. This has most notably been the case since the mid-2000s with the increasing global media attention devoted to the issue.

In the period following the rejection of the Kyoto Protocol by the United States in 2001, and the antipathy to climate action evident from the Bush administration (and later the Trump administrations), the COPs had a very important role in keeping climate change on international and domestic political and popular agendas. In the wake of increased familiarity with the climate issue in the last decade, and the heightened political priority attached to it, this function has declined in importance. Nevertheless, even now, the weeks around a COP are guaranteed to see significantly enhanced media and other discussions of climate change. Moreover, the emerging populist backlash against climate action in numerous countries may see this function take on an important role once again.

d) The COP is a "climate jamboree"

As observer participation has grown over time, COPs have become "the place" for the world's climate community to meet, to network, to exchange ideas and to be seen. Much of this activity centres around organized events and workshops, but much is also free flowing. The coffee lounges and the pavilions of a COP are hotbeds of discussion and networking.

It may be time to ask questions about the scale of this function. With participation in COPs surpassing 40,000, and even 80,000 at COP28 (Dubai, 2023), one cannot but wonder if such a jamboree, and its associated greenhouse gas emissions, can be justified, especially when measured against the very modest impacts of COP outcomes.[6] Flying to an international meeting is not the only way to demonstrate engagement, and it is definitely not the most impactful.

e) The COP is a mobilization opportunity

In recent years COPs have become important opportunities for governments and others to announce initiatives designed to reduce emissions and promote adaptation, thereby building momentum towards a more climate-secure future. The initiatives have been varied, covering multiple sectors or approaches, ranging from low-emission vehicles to methane reduction, from reforestation to coal phase out, and from solar alliances to clean energy research and development, as well as adaptation and finance pledges. They have involved governments at various levels, but also the private sector and civil society; many have been announced by heads of state or government. While modest prior to 2015, this trend intensified at COP21 in Paris and has been significant ever since, especially at COP26 (Glasgow, 2021) and COP28 (Dubai, 2023).

Some commentators have lamented this development of "on the side" initiatives and announcements as detracting from the formal negotiations. I disagree. The negotiations have been declining in importance and impact. The emergence of these "real world" action initiatives has meant that COPs can contribute to changing economic and social realities and be more than "talk-fests" that adopt "decision" texts heavy on rhetoric and the future and short on real action and impact.

---

6. Approximately 2,000 delegates attended COP1 in 1995. The next dozen COPs saw participation numbers well below 10,000, except for COP3, which did involve about 10,000. COP15 in Copenhagen (2009) and COP21 in Paris (2021) saw spikes in participation to almost 30,000, and recent COPs have attracted even greater numbers. See: https://unfccc.int/process-and-meetings/parties-non-party-stake holders/non-party-stakeholders/statistics-on-non-party-stake holders/statistics-on-participation-and-in-session-engagement.

This does not mean that the "on the side" initiatives are without problems. Too often, they are ambitious in intent but loosely defined. The follow-up, and tracking or accountability mechanisms, are almost always weak and non-transparent, and implementation tends to lack the vigour of the announcement. Nevertheless, these mobilization initiatives, if properly designed and tracked, have the potential to drive fundamental economic and social change (see Section 3.5b below).

3.3. A Thematic Chronology of the COPs

The negotiating history of COPs can be divided into four phases:

a) 1995–2001: Full-Scale Negotiations

COP1 (Berlin, 1995) launched the negotiation of what became the Kyoto Protocol, adopted by COP3 (1997). The president of COP1, Angela Merkel, then Germany's environment minister, played an instrumental role in achieving consensus on the hard-fought "Berlin Mandate" decision, demonstrating the skills that would later make her a successful and globally respected German chancellor. The negotiations in this period, masterfully led by Ambassador Raúl Estrada Oyuela, were focussed on achieving agreement on the Kyoto Protocol; numerous housekeeping and organization-building matters were nevertheless addressed as well.

The years following COP3 saw intense and detailed negotiations on the rules to govern the implementation of the Protocol, with a particular focus on the groundbreaking development of systems for emissions trading and related market mechanisms. The process wrapped up at COP7 (Marrakech, 2001) with the adoption of a detailed "rule book." However, challenges did emerge. COP6 (The Hague, 2000) ended in failure. When it resumed in summer 2001, governments adopted the "Bonn Agreements" in what can be interpreted

---

Kyoto Protocol – An Agreement for Its Time

I was the secretariat team leader in support of the Kyoto Protocol negotiations. Our team supported the chair of the Ad Hoc Group on the Berlin Mandate (AGBM), H. E. Ambassador Raúl Estrada Oyuela of Argentina, by organizing the negotiating sessions and preparing the documents, including the drafts of the treaty. Ambassador Estrada is the "hero of Kyoto" for his skillful leadership of a very difficult negotiation over two years and for inspiring, cajoling and bullying delegates toward agreement. His chairing of the last night of negotiations in Kyoto, live on the web, was masterful. The AGBM unanimously recommending the text of the Kyoto Protocol to COP3 for adoption was, for me, after days without sleep and the risk of failure ever-present, the most emotional moment of my UNFCCC career.

---

as the "rest of the world" coming together to move forward despite the US government's rejection of the Kyoto Protocol.

b) 2002–2010: Spinning Wheels

This period was marked by a continuing lack of consensus on how to move forward, especially in regard to the nature of future undertakings and differentiation between the commitments of industrialized and developing countries. COP15 (Copenhagen, 2009) marked a crisis when negotiations relating to a new agreement and the continuation of the Kyoto Protocol collapsed. Heads of state and government, including those from the United States, China, and India, agreed on a short and "punchy" outcome statement. However, because of the secrecy and lack of inclusiveness in its negotiation, the "Copenhagen Accord" was not "adopted" but "taken note of" by the COP, and the COP erupted in hours of acrimonious debate.

Copenhagen – Successful "Failure"

COP15 in Copenhagen was a remarkable experience in many ways:

High expectations met low political will and profound disagreements.

A COP president resigned in-session and was replaced by the prime minister of her country. Expecting a carefully choreographed success, Prime Minister Rasmussen stumbled undiplomatically through a disastrous closing of the conference.

Hours of recriminations in the closing plenary (I advised presidents/vice presidents who presided in succession through the drama).

The "Copenhagen Accord" was negotiated by a group of heads of state/government but found illegitimate by the rest of the COP and barely taken note of by the conference. It is now seen as the most successful failure in COP history (i.e., substantive progress but failed process).

Organizational highs and lows over 100 heads of state/government attending and speaking (a first!). But also hundreds of delegates forced to line up in the cold and snow for hours due to bottlenecks and an overfull facility.

The process failures at COP15 weakened trust and meant that in future small group negotiations to broker a final COP outcome became unacceptable and alternative methods of negotiation needed to be used.

Through its effective diplomacy, the Mexican presidency of COP16 (Cancún, 2010) was able to take the key points from the Copenhagen Accord (e.g., temperature change thresholds, funding goals) and turn them into legitimate, consensus COP outcomes, which, in due course, found their way into the Paris Agreement. Despite ongoing differences and very modest COP results, the period did see a critical milestone— the entry into force of the Kyoto Protocol (2005).

c) 2011–2021: A New Consensus Emerges

Paris – Political Will, the Corrections, and the Euphoria

The Paris Climate Conference (COP21, 2015) was the ultimate climate COP, a substantive and procedural success infused with ambition, goodwill and promise. The French COP presidency under then Foreign Minister Laurent Fabius brilliantly guided the process over two years, serving with neutrality, political savvy, diplomatic astuteness, and flair. As a secretariat, it was a dream experience to work with and support such a professional presidency in a true partnership. It meant a negotiation, and a result, of which all governments could feel ownership. Reading into the record in the COP plenary the "minor editorial corrections" to enable adoption of the Paris Agreement, especially one significant substantive blooper that had crept into the text, remains one of the high points of my COP experiences.

The impacts of climate change were becoming more apparent and the pressure for agreement on actions increased. In parallel, governments were able, slowly and incrementally from COP to COP, to find consensus on concepts that would allow a new agreement to be achieved.[7] This process culminated

---

7. Particularly notable points in this consensus-building process were:
   - The adoption of the Durban "Platform" decision (COP17, 2011), which launched a process to develop post-2020 a "protocol or another legal instrument."
   - The adoption of a second commitment period for the Kyoto Protocol (COP18, 2012). While not particularly impactful in terms of emissions limitations, it was an importantly symbolic demonstration by industrialized countries of their acceptance of the leadership prescribed in the Convention thereby clearing a hurdle and allowing further discussion of the nature of commitments.

in COP21 (Paris, 2015) adopting the Paris Agreement, an agreement more ambitious and comprehensive than most had expected, although many of its commitments were "soft," relating to conduct rather than results.

What followed was a further period of negotiations to develop the rules for implementation of the Paris Agreement, which was largely completed at COP26 (Glasgow, 2021).

d) 2022–? *"Crisis? What Crisis?"*

While evidence of the declining relevance of negotiations had been emerging before COP26, questions about the purpose and impact of COPs have grown louder in recent years. The fundamental question relates to the purpose of COPs in a world where:

- the necessary systemic rules have already been adopted;
- foundational international treaties exist and new multilateral climate agreements are not envisaged; and,
- action by governments and others does not depend on further multilateral agreements.

The challenge now and moving forward is to *implement nationally* what has been agreed internationally, a task at which governments and the private sector are failing abysmally. It is hard to conclude that COPs have "shifted to implementation mode" when their adopted decisions remain heavy on rhetoric and long-term targets or signals and short on elements that drive action on the ground.

3.4. Approaching COP30

When I began my career in the UNFCCC Secretariat in 1993, little did I imagine that three decades hence I would be writing

---

- The United States and China—the G2—worked to find concepts and wording that could bridge differences. Progress was captured in the outcomes of COP20 (Lima, 2014).

commentaries on climate negotiations or, even more starkly, that the global response to the climate challenge would have fallen so short of what was needed. With global emissions now more than 60 percent higher than in 1990, and with global average annual temperature increase for 2024 breaching the 1.5°C threshold,[8] the shocking failure of governments and other responsible actors is all too evident. Where does this leave the multilateral climate regime as COP30 approaches? A number of observations can be made.

The tendency to see COP30 as a "save the world' moment must be avoided at all costs. Such "last chance to save the world" hysteria has been used to describe previous COPs. Such rhetoric reflects a fundamental misunderstanding of the role and potential of COPs. COPs are, after all, only one part of the international response to the climate emergency; and this international response is itself only part of the broader socio-economic effort involving governments at all levels, the private sector, civil society, and even individuals.

Setting aside the hype, the role of COPs is fundamentally to help governments act domestically on climate change by agreeing at the international level on directions and principles of cooperation. COPs have, to date, done this rather successfully, having delivered two virtually universal treaties under the Convention, clear goals, a comprehensive system of rules and principles for implementation, and institutional architecture to facilitate cooperation. The onus is now on governments and the private sector to act on implementation within the parameters established by the international regime. They are failing to do so: The nationally determined contributions (NDCs) under the Paris Agreement are woefully inadequate, as are governments' policies and actions to implement the NDC targets and, even more so, to meet the

---

8. Global average annual temperature change measured against a pre-industrialization average.

goals of a 50% reduction in global emissions by 2030 and "net zero" by around 2050.[9]

It is unrealistic to think that COP30, or indeed any COP, will deliver the miracle turn around. Governments and the private sector need to get serious about urgently reducing emissions, preparing for adaptation to the increasingly evident impacts of climate change, and, in the case of industrialized countries, supporting developing countries to do both. Politicians and corporate leaders need to stop playing political games and put in place the solutions that are readily available and can be designed and implemented equitably and fairly. The idea that adopting COP decisions calling for 50 percent global emission reductions by 2030 constitutes progress, while global emissions are still rising, strains credulity.

This is not to suggest that COPs are a waste of time and effort. They are a tool to help governments strengthen collective actions, reinforce domestic policies where necessary, take stock, and advance implementation in a wide range of areas. In this context, several recent steps are important, such as progress on "loss and damage," the next round of NDCs, adaptation, and a signal to "transition away from fossil fuels."

The question does arise as to whether more can be expected from the COPs and the international climate regime beyond the norm.

3.5. How Could COPs Help Governments Be More Ambitious?

Business as usual at COP30, and beyond, will not change the trajectory of global emissions nor increase support for adaptation actions. If governments do, however, decide that they really want to enhance action on climate change rather than

---

9. New NDCs are due in early 2025. They should be carefully scrutinized to see if governments, especially among the G20 countries, are enhancing their actions to reduce emissions and thereby match their promises.

just talk about such enhancement, COPs and the international climate change regime may be able to help them. This section will explore options that could be considered.

In elaborating possible options, one must accept certain realities. These include:

- Governments have traditionally rejected strong compliance regimes in most international agreements, and emphatically so in relation to climate change.
- New universal agreements with binding commitments applicable to both industrialized and developing countries are a non-starter due to the deep-seated differences between North and South with regard to "responsibility" and financial support, not to mention the UNFCCC process requirement for consensus and the reality of radically differing national circumstances (e.g., energy mixes). Similarly, an agreement limited to only industrialized countries is also out of the question.
- The COP process and its negotiators are very conservative in their attitudes towards what a COP can or should do. This has limited the possibility of innovation and meant that process reform has never progressed. Witness the dismal situation regarding issues such as agenda modernization, virtual versus in-person meetings, process explosion versus process streamlining, and so on.

This means that thinking outside of traditional COP parameters will require those governments that are serious about using multilateralism to up their game, be creative, and push very hard.

a) Transparency and accountability

The Convention and the Paris Agreement establish a fairly robust system of transparency and, hence, accountability through the rules on reporting and review of national actions. Nevertheless, they are "facilitative" and non-confrontational and, hence, generally benign. One of the simplest options to promote ambition would be to strengthen the transparency regime with a focus on NDC implementation. While this might provide some incremental peer pressure on governments, and may be worth pursuing for that reason alone, its ability to promote more ambitious action would be limited unless it helped generate more domestic political pressure. A voluntary system (see below) could be a variant that would have somewhat more impact.

Transparency and accountability have also been a significant shortcoming in relation to the voluntary initiatives announced by some governments on the side of COPs in recent years. Strengthening and institutionalizing accountability in these arrangements would be straightforward and could deliver more results (see also next subsection).

b) NDCs and "policies and measures"

The principal shortcoming of Paris Agreement implementation is the absence of ambition in NDCs,[10] compounded by weak implementation of the undertakings made in the NDCs, especially by the G20 countries.[11] A number of avenues could be explored to address this lacuna. For example, the G20 countries, or other groups of countries, could come to political understandings to increase the level of ambition in their NDCs, either with regard to reduction targets or particular collaborative initiatives. However, the track record of such political undertakings and targets, absent some element of compulsion, is not encouraging.[12]

10. While there is an obligation to submit NDCs, they constitute non-binding pledges of action or target-achievement (Paris Agreement, Article 4.2).

11. The world's 20 leading economies (G20) are responsible for, and have the ability to manage directly, approximately 80 percent of global emissions.

12. It must be acknowledged that the strengthening of NDCs will come not from international pressure but from domestic lobbying and mobilization and requires political will to move from "business-as-usual."

Of more interest could be sector-specific agreements among limited sets of committed countries to take particular action. These have precedents in the form of earlier informal conversations (e.g., workshops) on particular policies and measures that could be implemented cooperatively. In more recent years, there have been numerous initiatives announced on the side at COPs, often at heads of state/government level, under which the governments and business sectors involved have made undertakings to take particular actions. Examples of these include an international solar alliance at COP21 in Paris, initiatives on zero-emission cars and vans at COP26 in Glasgow, and efforts to cut methane emissions at COP28 in Dubai.

Such agreements on policies and measures or sectoral action would be a means through which key governments and businesses could cooperate in undertaking specific actions, thus generating momentum towards changing the overall economic landscape. Initiatives of this sort would need to go beyond previous voluntary pledges and be actual agreements among the participants, with reporting obligations, a secretariat tracking mechanism, and some form of legal obligation. These need not be universal agreements under the UNFCCC but agreements among groups of countries announced and, to the extent possible, tracked within the COP system. Given the emission realities in the fossil fuel and transport sectors, these could be a priority.

As part of the COP's awareness-raising function, opportunities could be developed to shine a spotlight on success stories in implementation (e.g., high-impact policies and measures) or on best practices. This could involve work by the UNFCCC subsidiary bodies or constituted bodies leading

---

In this context, several recent cases where citizens groups have brought legal action against governments in national courts, drawing on both domestic and international laws, have shown some promise and should be further pursued.

to authoritative reports. Moreover, interested governments could promote discussion, either in formal settings or on the side, of key concepts, such as eliminating fossil fuel subsidies or phasing down/transitioning away from coal/fossil fuels (use, exploration, and extraction).

In addition, there exists a heretofore unused provision of the Convention that could be used to help strengthen credibility and rigour in initiatives. Specifically, Article 7.2(c) of the Convention enables two or more Parties to ask the COP to "facilitate...the coordination of measures."[13] This provision could lead to action in UNFCCC bodies and the financial mechanism, as well as facilitate engagement by the Secretariat (e.g., in reporting or awareness-raising).

There are two additional elements that need to be mentioned. First, industrialized countries (through their bilateral development assistance) and multilateral development agencies (via their programming and lending) should become serious about supporting developing countries in formulating meaningful NDCs and, even more importantly, in implementing them. Secondly, the systems of accountability, or reporting and review, could be made more rigorous. While it would be difficult to negotiate a ramped up universal system of international expert review, with real assessments of the level of effort and implementation, a voluntary system whereby leading countries could submit themselves to such reviews could help to build credibility and inspire not only action and ambition among the committed but provide peer pressure for other countries to join. Such a system could be established by a COP decision or perhaps by using the above-mentioned Article 7.2(c).

---

13. Convention Article 7.2: The COP shall (c) "Facilitate, at the request of two or more Parties, the coordination of measures adopted by them to address climate change and its effects, taking into account the differing circumstances, responsibilities and capabilities of the Parties and their respective commitments under the Convention."

c) *Finance, development, and investment*

As mentioned above, support for developing countries to reduce emissions (as well as to adapt to climate change impacts, and to address loss and damage) is an essential element of seriously meeting the climate challenge. The prospects of significantly improved performance by contributor countries in meeting their general obligations to assist developing countries are not encouraging. Nevertheless, support can have a significant impact, especially for the smaller economies seeking to transition their energy systems or to implement sustainable development strategies. The onus must be on the G7 and the G20 countries to lead in this regard, especially in changing investment patterns and prevailing development models. The multilateral development banks and international financial institutions also need to be part of the solution to ensure development without fossil fuel investments. The same applies to the commercial financial world.

The ability of COPs to formally influence these discussions is very limited. Nevertheless, as high-profile events involving heads of state/government, COPs can exert some pressure. The engagement of heads of state/government and ministers of finance and development is indeed key, as is a complete mindset change on their part. Rhetoric must shift into actual actions.

Governments have been loath to seriously consider targeted international initiatives to generate funding, such as various options for taxes, levies, and charges that would generate revenues for development and investment. The time may be ripe, however, to put these options back onto the agenda. Carbon taxes, if fairly designed and implemented, can not only internalize environmental costs and influence economic behaviour but also generate significant revenue, even at fairly low levels. Other examples often discussed but never agreed upon include levies on air travel and cargo or on financial services, to name two examples. While there may be some role for COP decisions in this area, agreements in other institutions and processes would be required, as would leadership by the major economic powers.

d) All of multilateralism

Those actively engaged in COPs tend to see the UNFCCC process as the centre of all climate action. This is a fallacy. Responding to the climate emergency requires an all-of-government and all-of-multilateralism approach. COPs can serve as a forum for general overview or advocacy, but in several sectors, the real responsibility rests with other international institutions and processes, most of which say the right things but do not deliver. Processes, such as those for civil and freight aviation (ICAO), maritime transport (IMO) and agriculture (FAO) need to become more than apologists for the corporate interests that dominate them. This will require governments to confront policy decisions domestically as part of all-of-government climate action and implementation, and then to break prevailing international inertia through leadership initiatives within the sectoral organizations.

e) Economic instruments and trade measures

New economic instruments and trade measures have received little multilateral attention, especially as regards their potential to enable greater ambition in climate action. One example would be coordination on the design and implementation of carbon taxes and other market mechanisms. The Convention is specific in its pro-trade orientation.[14] This, however, does not preclude negotiations on this topic, either among all governments or a subset. Neoliberal trade rules and policies have contributed to the climate emergency by facilitating industrial and emission shifts. National climate ambition should not be frustrated by unsustainable trade rules designed for

---

14.   See Convention Article 3 (Principles), para 5.

another purpose. International cooperation on border tax adjustments, for example, could be facilitated through the COP to prevent abuse. The COP process, after all, has considerable experience in agreeing on detailed rules relating to emissions trading and emission offsets (Clean Development Mechanism).

3.6. Conclusion

As COP 30 approaches, it is appropriate to celebrate the significant achievements of climate multilateralism—three major and innovative treaties and an elaborate system of institutions and processes to assist and encourage governments in tackling the climate emergency. Nevertheless, the reality is that most governments have failed to implement the letter and spirit of the climate treaties and, in many cases, even to fulfill their basic legal obligations. While some may argue that rising global emissions and overshooting global temperature-change thresholds mean that multilateralism has failed, the reality is more nuanced. In my view, multilateralism has done what it can be expected to do, given the realities of state behaviour.

The onus is now on committed governments to act—foremost those in the G20. Governments need to raise their levels of domestic ambition and build international coalitions of like-minded countries to drive action forward. Only in this way can economic and social realities be changed and climate laggards be forced to join a new, more just, more prosperous, and more sustainable reality. Waiting for additional universal agreements constitutes dangerous procrastination.

COP30, and subsequent COPs, can help move the world in the right direction. They will do so not through the negotiation of lofty COP decisions with far distant goals, but by advancing pragmatic and targeted cooperation and initiatives to help governments reduce emissions now, adapt to growing climate change impacts, and enhance support to developing countries in these efforts.

A Fair, Funded, and Fossil-Free Future: A Parliamentarian's Role in International Climate Action

Rosa Galvez

In my native country of Peru, I grew up surrounded by the vast landscapes of the Andes Mountains, in proximity to the Amazon Forest, where we can find some of the most distinct biodiversity in the world. Peru, known for its natural marvels, was where I discovered early on the wonders of nature and of biodiverse habitats. From a young age, I was blessed with experiences that helped shape my vision of the world as I see it today and my understanding of the importance of nature for humanity to thrive. These experiences taught me that nature is not something to be conquered or taken for granted, but rather respected and understood. I saw firsthand how people's relationship with the environment can sustain them, but also how it can be threatened by unsustainable practices.

The foundation for my deep connection to nature and the balance between tradition and progress was laid by my grandparents. My maternal grandfather, who worked as an architect and engineer in Peru, built homes and fostered community. I vividly remember visiting housing construction sites with him as a teenager, where he supervised well drilling for drinking water. On the other hand, my paternal grandmother was a "wise woman"—a custodian of ancestral knowledge passed down through generations. She could neither read nor write, but she had an encyclopedic understanding of traditional knowledge. I hiked the Andes with her into the Amazon jungle in search of medicinal plants. I also assisted her during the miraculous moments of childbirth, further deepening my reverence for life and the natural processes that sustain it. She taught me that nature holds the key to human survival and well-being—lessons that could not be found in books, but in the world around us.

I loved and admired both my grandparents. They showed me, through their own unique ways of knowing the world, that we must seek equilibrium between learning from nature and development at its expense. The progress of society must not come at the expense of the environment, and conversely, respecting nature does not mean rejecting innovation. Their influence instilled in me the belief that true development must honour both tradition and the future.

I became aware of the global environmental crisis at a young age. I remember being shocked by images of air pollution in major cities across the world. These scenes made a lasting impression on me, and I envisioned myself creating mechanical tools to clean air and oceans. I longed for cleaner air, water, and soil for all humans to enjoy. It was for that reason that I became an engineer with a particular interest in cleaning up our environment. Whether it was the rich landscapes of the Peruvian Andes or the vast natural beauty of my adoptive country, Canada, I have always felt a deep connection to our planet's natural ecosystems. This connection has been the foundation of everything I do, guiding my path as both an engineer and an environmental advocate.

Nowadays, I have the utmost privilege of advancing environmental policy as a Canadian senator at the Parliament of Canada. My role allows me to contribute to shaping the future of environmental stewardship, both in Canada and at crucial international gatherings for the environment such as the Conference of the Parties (COP) of the United Nations Framework Convention on Climate Change (UNFCCC). I am fortunate to work alongside leaders from around the world to address the pressing challenges of climate change, biodiversity loss, and environmental degradation.

My journey began in the tall mountains and lush forests of Peru, and it has led me to a position where I can help shape the future of environmental policy and advocate for the basic necessities of clean air, water, and soil. But no matter how far I travel, I carry with me the lessons of my grandparents: the need for balance between development and nature, and the understanding that the health of our planet affects us all. My mission remains clear: to work tirelessly for a world where pristine environments are rights, enjoyed by all, and where the natural beauty of our planet is preserved for generations to come.

4.1. My Introduction to COP

My first experiences with the UNFCCC's COPs were as a researcher in environmental engineering. Early on, I had decided that I wanted my research to be centred around decontamination and environmental restoration, and I had a particular interest in aiding and enhancing the natural restoration and remediation processes of nature. Therefore, my main focus as a researcher and engineer was the environmental impacts of infrastructure projects, such as transportation, housing, energy production, waste, and water systems, and how our changing climate is affecting these essential aspects of human society. My interest in learning more about the technical indicators of environmental health led me to attend my first COP and for most of my career, I engaged sporadically in these international gatherings, staying informed on any development coming out of these meetings and integrating relevant scientific research into my teachings.

I attended my first COP, COP20, in December 2014. COP20 took place in my hometown of Lima, Peru. I was particularly proud to finally engage in an UNFCCC conference in the country that first steered my attention towards environmental issues. As an attendee of COP20, I had the opportunity to attend preparatory meetings and was an engaged observer from the scientific community. Although I did not know it then, this would become a significant moment for the international community and for my own journey.

The Lima conference's main objective was to create a draft negotiating text for a global climate agreement that would lead the world in the fight against climate change. It laid the groundwork for an international framework that would guide every country for years to come. Lima became the key stepping-stone towards what would eventually become the Paris Agreement, which was signed at the following COP in 2015.

The draft text was crucial for defining how countries would commit to limiting global warming and how the international community would cooperate in addressing climate change. The text acknowledged one of its objectives as the achievement of "low greenhouse gas climate-resilient economies and societies, on the basis of equity and in accordance with their historical responsibilities."[1]

The text also acknowledged the principle of common but differentiated responsibilities among party countries "in order to achieve sustainable development, poverty eradication and prosperity for the benefit of present and future generations of humankind, taking fully into account the historical responsibility of developed country Parties and their leadership in combating climate change and the adverse effects thereof, and bearing in mind that economic and social development and

1. UNFCCC, "Elements for a Draft Negotiating Text," Agenda Item 3 (COP20, Lima, December 10, 2014), https://unfccc.int/files/meetings/lima_dec_2014/in-session/application/pdf/adp2-7_3_10dec2014t_np.pdf.

poverty eradication are the first and overriding priorities of developing country Parties."[2]

This conference encouraged countries to start preparing their intended nationally determined contributions (INDCs), which were their individual commitments to reduce greenhouse gas emissions. These INDCs would later form the backbone of the Paris Agreement and be transformed into binding nationally determined contributions (NDCs) after the adoption of the agreement.

COP20 also emphasized the need for developed nations to provide financial support for climate mitigation and adaptation efforts in developing countries. There were calls to bolster the Green Climate Fund (GCF) to help vulnerable countries deal with the effects of climate change.

Another important theme was strengthening the global response to climate adaptation and addressing loss and damage from the impacts of climate change, especially for countries most vulnerable to climate change impacts. As someone who grew up in a developing country, I feel particularly strongly that richer countries need to do more to help vulnerable countries deal with the financial loss and damage caused by climate catastrophes, as well as provide funds to help those countries mitigate and adapt to the consequences of climate change.

The final outcome of the Lima COP was a Call for Climate Action, which provided the foundation for negotiations that continued into COP21 in Paris. The document outlined guidelines for countries to submit their INDCs and clarified the process for reporting and reviewing these pledges.

The Lima meeting was considered critical for securing the necessary political momentum and negotiating framework leading up to the Paris Agreement, but there were challenges in balancing the responsibilities of developed and developing nations regarding emission reductions and climate finance.

---

2. UNFCCC, "Elements for a Draft Negotiating Text."

Indeed, the Lima conference not only gave me hope for a prosperous future for all but also helped shape my future engagement with the UNFCCC COPs. Nevertheless, as I participated as a scientist at the Lima conference, I had limited influence and ability to change the path of negotiations. However, I would later engage in a much different role with a larger focus on policy when I attended COPs as a senator.

4.2. Nomination to the Senate

In 2016, I became an independent senator at the Senate of Canada. There, I used my expertise in environmental engineering to advance meaningful environmental and climate policy in Canada. From the outset, I recognized the urgent need to address climate change on both national and international levels. My work in the Senate allowed me to advocate for stronger domestic policies to protect Canada's natural landscapes, reduce emissions, and promote sustainable development, while also sharpening my focus on the importance of international cooperation for environmental protection.

I was unfortunately unable to attend COP23, which took place in Bonn, Germany. The main goal of COP23 was to make progress on the Paris Agreement's implementation rules (often called the Paris Rulebook) to ensure that countries would have a clear framework for meeting their climate commitments by 2020.

One of the key outcomes of COP23 was the launch of the Talanoa Dialogue Platform.[3] Talanoa "is a traditional word used in Fiji and across the Pacific to reflect a process of inclusive, participatory and transparent dialogue,"[4] and the Talanoa Dialogue Platform is a platform for countries to assess collective progress towards climate goals. It focused

---

3. "2018 Talanoa Dialogue Platform," UNFCCC, accessed January 29, 2025, https://unfccc.int/process-and-meetings/the-paris-agreement/the-paris-agreement/2018-talanoa-dialogue-platform.
4. "2018 Talanoa Dialogue Platform."

on raising ambition and providing support for vulnerable nations like Small Island Developing States, which are disproportionately affected by climate change.

By COP23, I, along with many others in the global community, had become keenly aware of the injustice that exists with respect to the disproportionate negative impacts of global warming. Indeed, small island nations, and the Global South as a whole, have contributed the least to total greenhouse gas (GHG) emissions, and yet they suffer the greatest and most intense negative impacts related to climate change, including soil and coastal erosion, sea-level rise, and changes in rainfall patterns. Prior to COP23, it was not evident to me that these divergent contributions to the climate crisis, and thus distinct responsibilities to pay the costs associated with adapting to a warming planet, were wholly recognized. Seven years after the need to support vulnerable nations was acknowledged at COP23, we continue to discuss the need for the Global North to take responsibility based on their "fair share" and there are processes and practices in place, like the nationally determined contributions, that aim to take into consideration the notion that our climate commitments can only be met if we recognize that each nation has a responsibility to act but that the Global North has a responsibility to provide assistance to the Global South and small island nations.

Unfortunately, the whole process of compensation and financial assistance to support the Global South is lagging where we need to be to achieve global climate commitments. As a Peruvian-born Canadian senator, I believe it is my duty to advocate for the rights of marginalized peoples and communities, not just nationally but also globally, and to foster meaningful dialogue between the Global North and Global South by bridging the gap between their often-conflicting perspectives.

The following COP, COP24, took place in Katowice, Poland, in 2018. Although I was unable to attend this COP, I followed the news out of the gathering with great interest.

The primary task of that conference was to finalize the Paris Rulebook, providing guidelines on how countries should implement their commitments under the Paris Agreement. COP24 was particularly significant because it was during this conference that the Climate Package was adopted. The Climate Package is a comprehensive set of technical rules and guidelines detailing the rules on transparency, mitigation, and adaptation, and setting standards for how emissions reductions should be measured and reported. In addition to the technical aspects, COP24 highlighted the critical importance of climate finance, particularly for developing countries. The discussions stressed that developed nations have a responsibility not only to reduce their own emissions but also to support vulnerable nations in coping with the impacts of climate change. This support includes financial resources, capacity building, and technology transfer. The conference called for enhanced commitments from developed nations to fulfill their climate finance pledges, particularly by reaffirming the goal of mobilizing $100 billion annually by 2020 to assist developing countries in their transition to low-carbon economies and to bolster their resilience against climate-related disasters.

Support for developing countries is a principle that resonates directly with me in my position as a senator. It is my role to represent my province, but also to protect the underrepresented and the most vulnerable. It is evident that many nations have less capacity to manage the impacts of climate change and to quickly transition to a low-carbon economy. This is also true within Canada, a large nation in which each province and territory is vastly different; geographically, economically, and politically. For example, the province of Quebec, as a long-time leader in renewable energy, already has a head start in the transition and on reducing its carbon footprint, while the economies of other provinces, like Alberta and Newfoundland, are more entrenched in fossil fuels. Domestically and internationally, we must consider each region's economic and technical feasibility to transition. Many nations have great potential for renewable energy development, but may require financial and technical support. Diverting the massive subsidies that continue to bolster the fossil fuel industry towards renewables, both domestically and internationally, is a low-hanging fruit that could easily shift the tide in climate finance and increase the flow of funds to support the most vulnerable.

4.3. COP as a Parliamentarian: A New Experience

After a brief hiatus, I once again started attending COPs, this time as a parliamentarian. As a Canadian parliamentarian, I became a member of ParlAmericas, a network of parliamentarians of the Americas, whose main pillars of cooperation include climate change. In 2021, I started serving as president of ParlAmericas' Parliamentary Network on Climate Change and Sustainability (PNCCS), which aims to promote "parliamentary diplomacy on climate action within parliaments, aligned with existing international frameworks that work towards combating climate change and achieving sustainable development."[5] It was created soon after the adoption of the Paris Agreement.

The PNCCS offers an Americas-wide platform for parliamentarians to exchange knowledge and policy ideas in collaboration with expert practitioners and civil society. The network includes all the Americas, including the island nations of the Caribbean, whose entire lands are particularly impacted by the rapidly changing climate. For the network, it has become apparent that mitigation of, and adaptation to, climate change is more than the need to protect our planet; it is an imperative to protect our livelihoods and those of the generations to come.

5. "About the PNCCS," ParlAmericas, accessed September 2024, https://parlamericas.org/about-the-pnccs/.

It was in 2019 that I officially participated in a COP for the first time as a parliamentarian. I was a delegate representing ParlAmericas at COP25, held in Madrid, which offered my first glimpse as a policy-maker. Attending this COP as a senator was a completely different experience compared to the experience I had attending as a scientist, and it opened my eyes to the complexities of international climate negotiations and the importance of parliamentary involvement in shaping climate commitments. I started to notice the geopolitical implications of these negotiations and especially how political games were delaying the much-needed climate action laid out by scientists. It is at this COP that I heard two contradictory speeches: one from scientists calling for action and warning about present and future dangers, and another from the passive political class pressured by the ever-present oil and gas lobby pushing their "deny, doubt, delay, and disinform" messaging.

The major focus of COP25 was to finalize Article 6 of the Paris Agreement, which deals with global carbon market mechanisms and how countries can trade carbon credits to meet their climate targets. Unfortunately, no consensus was reached on Article 6, leading to a deadlock that was carried over to future COPs. The conference underscored the urgency of raising climate ambition, though there was a general disappointment over the lack of substantive progress.

Personally, having not attended COPs for many years prior to COP25 in Madrid, I was utterly disappointed to witness the significant presence of corporate sponsorship throughout the conference. I was particularly shocked at the massive presence of lobbyists and the often-questionable corporate sponsorships helping to fund the COP, something I had not noticed as much when I was a scientific observer. These multinational corporations included the mining sector, oil and gas, and power utilities, whose own activities fuel climate change. Such exposure at the most important climate change gathering in the world gives these polluting corporations an international platform for greenwashing and gives them significant influence at the negotiations.

At COP25, it became increasingly apparent that corporations, particularly those with vested interests in fossil fuels and other environmentally harmful practices, had a notable influence on the proceedings. One glaring example was the overwhelming visibility of promotional materials, such as enormous banners advertising Coca-Cola. This sponsorship raised critical questions about the integrity of the climate negotiations and whether the presence of such corporations was undermining the urgent call for climate action.

The participation of oil lobbyists in discussions about climate policy seemed to be in contradiction of the intentions of these deliberations because their primary business models are based on the extraction and sale of fossil fuels—one of the main drivers of climate change. I felt disillusioned by their presence. In a turn of events that I could not have anticipated, the very entities responsible for significant carbon emissions were being allowed to shape the narrative around climate solutions, potentially diverting attention from the radical systemic changes needed to address the climate crisis effectively.

Following COP25, I have been acutely aware of the ever-increasing presence of lobbyists doing everything they can to weaken any international agreement. In fact, since COP29 in Baku, Azerbaijan in 2024, many environmental activists have been calling for the ban of fossil fuel lobbyists at the international negotiations, as the sole purpose of their presence is to create barriers to real progress, as evidenced by the failure to explicitly reiterate the call to phase out fossil fuel subsidies in the COP29 agreement, despite this having been clearly stated in the previous COP agreement.

Since COP25, I have been a Canadian and ParlAmericas delegate at COP26 in Glasgow, Scotland, at COP27 in Sharm el-Sheikh, Egypt, and at COP28 in Dubai, United Arab Emirates. These conferences have been critical learning experiences, where I have gained valuable insights into the intricacies of international climate frameworks and the collaboration required to achieve global goals.

As a parliamentarian, my interventions at these COPs have mainly been to audiences of other parliamentarians. My goal is to deepen their understanding of the interconnectedness that exists between the environment, the economy, and society. My impression has been that, given my engineering and scientific background, parliamentarians listen intently to what I say about climate finance and the impacts of climate change on infrastructure and society. Indeed, my experience as a teacher has allowed me to simplify complex concepts and I believe this has also contributed to increasing my impact among my peers.

By participating in these conferences, I gained exceptional hands-on experience and deepened my understanding of international agendas on environmental matters. I have also expanded my involvement with a wide range of international agencies and civil society organizations. Among them is Parliamentarians for a Fossil-Free Future—a network of parliamentarians around the globe "calling for a full, fair, and funded phase-out of fossil fuels and transition to renewable energy systems."[6]

4.4. COP26: A Turning Point for Climate Finance

On a more positive note, the most interesting COP I attended was COP26 in Glasgow, Scotland. COP26 was particularly significant, as it was initially supposed to take place in April 2020 but was postponed due to COVID-19. COP26 was, therefore, the first chance world leaders had to convene and discuss the pressing subject of climate action since the beginning of the COVID pandemic. It was also the deadline for countries to update their NDCs and commit to more ambitious climate action.

---

6. "Homepage," Parliamentarians for a Fossil-Free Future, accessed September 2024, https://www.fossilfuelfreefuture.org/.

I started paying attention to the need to push the finance sector to become part of the solution in the transition to a low carbon economy. By this point, I had spent two years assessing the situation and had concluded that Canada was lagging in the transition, with the vast majority of investments being injected into the oil and gas sector in particular. More specifically, much of Canada's investment focus was on the oil sands. Thus, I was worried that Canadians were putting all our eggs in one basket. I was invited to speak at several events. For example, I opened the International Parliamentary Union's (IPU) Parliamentary Meeting at COP26 called "Expert Opinion Session, November 7, 2021 - A Clean Recovery with a Particular Focus on Climate Finance," where I spoke about climate finance and the need for a clean transition. As I reminisce on the speech that I gave, I find that the following paragraphs are still very pertinent today:

> Several months into the COVID-19 pandemic, it became clear for many of us that governments across the world would have to inject unprecedented amounts of money to support its citizens and to recover from the socio-economic crisis provoked by the lockdowns. These lockdown conditions have taken a disproportionate toll on lower income individuals, the elderly, and groups who already bear a degree of structural oppression. Further, the pandemic has revealed a system wherein we exploit the finite natural resources of our planet with the illogical expectation of infinite growth, and governments subsidize environmentally destructive behaviour through support of polluting industries and corporations. Unbridled economic growth is the root cause of ecological destabilization.
>
> *As stimulus began to flow, I started reflecting on the ultimate goal and the most* efficient way to achieve it. These reflections led to the publication of a white paper entitled "Building Forward Better: A Clean and Just Recovery from the COVID-19 Pandemic" publicly available and in three languages, drawing inspiration from various organizations

advocating for a holistic approach to rebuilding Canadian society to achieve greater overall collective well-being by proposing specific public policies in great detail. It also addresses the funding of a clean recovery, where and how the funds can be recouped, and included a set of core policy recommendations. Although the white paper focused on the Canadian experience, these proposed policies, and the principles on which they are based are applicable across the world.

A clean and just recovery is one that puts people before profit and focuses on furthering and eventually achieving human and ecosystem wellbeing. Such a goal implies the development of principles and tools that not only ensure the costs and benefits of the recovery are distributed equitably, but also help shift our concept of growth to be centred around sustainable prosperity—after all, the economy must serve society, not the other way around.

The outcome of COP26 was the Glasgow Climate Pact, which aimed to keep the 1.5°C goal "alive" by recognizing the impact of human activities on global warming. The Glasgow Pact was meant to re-engage countries whose NDCs were falling short, encouraging countries to revisit and strengthen their NDCs by 2022. Article 6 was finally agreed upon, establishing global carbon markets and rules for trading emissions reductions. Countries further committed to phasing down unabated coal and reducing methane emissions.

COP26 was the first time I really understood the importance of climate finance in transitioning to a green economy. Particularly interesting to me were the calls to increase the commitment to $100 billion per year to support developing nations. I participated in several panels on this topic. I remember most vividly the Globe Legislators Summit, and the Inter-Parliamentary Union's panels, where domestic and international finance of loss and damage and just transition were hot topics. I specifically advocated, along with many fellow parliamentarians from across the globe, for increased climate finance ambition from developed countries as well as for the adoption of a more stringent legislative framework to guide the financial sector in its transition. This, unfortunately, is still a work in progress, as these proposals often face political barriers and opposition from financial institutions themselves.

4.5. Attending COP27 and COP28

In 2022, I attended COP27 in Sharm el-Sheikh, Egypt, which aimed to shift the focus from promises to concrete action, particularly for countries already experiencing the most severe impacts of climate change. The theme of adaptation came to the fore, as vulnerable nations called for greater support to cope with extreme weather events and rising sea levels. A major breakthrough at this summit was the establishment of a "loss and damage" fund to assist developing countries in coping with climate disasters. Climate finance was another critical issue at COP27, with increased pressure on wealthy nations to meet their promises, including the longstanding goal of providing $100 billion per year to support climate action in developing countries. Despite these financial commitments, many attendees expressed frustration over the slow pace of implementation, emphasizing the need for greater accountability and faster action to meet the goals of the Paris Agreement. The summit underscored the urgent need for more accessible and targeted funding for adaptation, especially for the most vulnerable nations.

In Sharm el-Sheikh, I continued my advocacy for stronger legislative guidance for financial institutions with fellow parliamentarians and with civil society. There was high hope to continue the progress on finance initiated in Glasgow the previous year. With the establishment of the loss and damage fund, there was a true feeling of progress and collaboration among nations.

However, COP27 was also notable for the presence of representatives from the fossil fuel industry, with 636 fossil fuel lobbyists in attendance. This was an increase from the 503 lobbyists at COP26 in Glasgow in 2021. This raised concerns about the growing influence of the fossil fuel industry at climate negotiations, raising questions about why there was such a strong presence of actors who are actively contributing to emissions. At the same time, the 300 or so Indigenous delegates present at COP28 were left feeling unheard and disillusioned.[7] Although COPs provide a platform for a wide variety of vulnerable voices, I feel that the overbearing influence of the fossil fuel sector eclipses those voices and muddies the debate around final agreements.

By COP28 in 2023, held in the United Arab Emirates, the number of fossil fuel lobbyists skyrocketed to a record-breaking 2,456. This marked the highest number ever recorded, sparking widespread criticism due to the strong presence of oil and gas representatives amidst calls for stronger climate action. These increases have led to major debates about the role of fossil fuel industries in climate negotiations, as many see their presence as a conflict of interest that undermines the goals of limiting global warming.

COP28 was expected to centre on a Global Stocktake of the Paris Agreement, a periodic process that assesses the collective progress towards the Paris Agreement's goals. COP28 was notable because it was the first time that a global stocktake would take place, which would assess how far countries have come in reducing emissions and meeting their climate targets. However, the outcomes remain uncertain due to significant friction and tension, as the summit's presidency was

---

7. Nina Lakhani, "Indigenous People and Climate Justice Groups Say Cop28 Was 'Business as Usual,'" *The Guardian*, December 23, 2023, https://www.theguardian.com/environment/2023/dec/13/indigenous-people-and-climate-justice-groups-say-cop28-was-business-as-usual; Dimitri Selibas, "Little Achieved for Indigenous Groups at U.N. Climate Summit, Delegates Say," *Mongabay*, December 19, 2023, https://news.mongabay.com/2023/12/little-achieved-for-indigenous-groups-at-u-n-climate-summit-delegates-say/.

held by the CEO of the national UAE oil company, which raised concerns about conflicts of interest.

There was increased attention on reducing emissions from sectors like energy, industry, and agriculture, as well as focusing on the transition to renewable energy. There were also some further discussions on strengthening carbon markets and climate finance. However, while the Green Climate Fund (GCF) underwent its second replenishment, I was rather disappointed that most climate-related finance issues were deferred to COP29. Furthermore, COP28 did not lead to the adoption of, or any decision on, rules for carbon markets, which creates uncertainty with regard to international carbon trading.

A highlight of COP28 was perhaps the agreement to a global transition away from fossil fuels, but this too fell short of the commitment that we ultimately need to make to phase out fossil fuels. Moreover, it is unfortunate that the COP28 deal also supports carbon capture and storage technology (CCS) and fails to recognize the limitations of CCS technologies. Lastly, I would note that I was thrilled to see the important role that Indigenous Peoples played at COP28 both through the COP28 Indigenous Peoples Dialogue with the UN Climate Change High-Level Champions, the COP28 presidency, and the UNFCCC Secretariat and throughout the conference at the Indigenous Peoples' Pavilion.

Looking ahead to COP29 in Baku, Azerbaijan, the focus was expected to shift to implementing the recommendations from the Global Stocktake conducted at COP28, enhancing climate ambition, and continuing discussions on loss and damage, finance, and adaptation. Anticipated outcomes included stronger commitments to increasing NDCs, clearer pathways for phasing out fossil fuels, and decarbonizing critical sectors.

Unfortunately, COP29 proved to be a disappointment for environmental advocates, as the final agreement failed to reiterate the call to phase out fossil fuel subsidies and negotiations were once again led by the president of a petro-state who described fossil fuels as a "gift from God." Many decisions have been postponed to a later gathering, causing further delays to meaningful progress. Meanwhile, we are already halfway through the decade that has been described as critical for climate action.

In 2025, COP30 will be held in Belém, Brazil. As a Peruvian-Canadian senator and president of ParlAmericas' PNCCS, I could not be more thrilled that a South American nation will host the next COP. Already, in anticipation of COP30, ParlAmericas has begun hosting a series of virtual meetings on diverse topics key to supporting the network's preparations for this pivotal meeting. With the return of this meeting to South America, I am hopeful that outcomes from this conference will be more ambitious and that the voices of the underrepresented and most vulnerable will be heard. Thinking back to my first COP in Lima, I am optimistic that I can attend COP30 in Belém with pride, and that negotiations at this COP will foster a more collaborative and holistic approach to climate action. I hope it will inspire us all to accelerate our efforts towards meeting our climate commitments.

4.6. Moving Forward: The Need for Greater Ambition

If the COPs described above had to be distilled to one central theme, it would be raising ambition. Unfortunately, while pledges have increased, we have not seen concrete action taken to release the promised funds to assist the Global South with the loss and damage caused by climate change. Similarly, domestic levels of funding to support adaptation and mitigation have fallen behind what was initially promised. Some nations, such as Sweden, Denmark, Germany, Norway, the United Kingdom, Costa Rica, New Zealand, and Finland, have made great progress in transitioning towards renewable energy. These countries are recognized not only for their commitments to net zero but also for the practical steps they are taking towards decarbonization, offering blueprints for others to follow. On the other hand, the majority of the Global North keeps increasing emissions. I am saddened to report that Canada has the worst record of all G7 nations,[8] and its plan to reduce emissions by 40%–45% below 2005 levels is insufficient.[9]

Each of the COPs had strong and optimistic slogans. Here are a few examples:

COP22 (2016 – Marrakesh, Morocco): "The COP of Action"
COP23 (2017 – Bonn, Germany, under Fiji presidency): "Further, Faster, Together"
COP24 (2018 – Katowice, Poland): "Changing Together"
COP25 (2019 – Madrid, Spain, under Chilean presidency): "Time for Action"
COP26 (2021 – Glasgow, Scotland): "Uniting the World to Tackle Climate Change"
COP27 (2022 – Sharm el-Sheikh, Egypt): "Together for Implementation"

Despite the ambitious goals set at each COP, the outcomes have often fallen short of the expectations needed to address the climate crisis. It is disheartening to see that, despite years of summits and strong pledges, promises have largely remained unfulfilled, with real-world action lagging behind. While emissions may be peaking, they have not yet been meaningfully reduced. The much-needed financing, pledged to help developing countries mitigate and adapt to the effects

8. John Woodside, "Canada's Embarrassing Climate Record Is Worst of G7 Nations," *National Observer*, June 1, 2021, https://www.nationalobserver.com/2021/06/01/news/canadas-climate-record-worst-g7-countries.

9. The Canadian Press, "Canada's Emissions Reduction Plan Falling Short: Environment Commissioner," *CTV News*, November 7, 2023, https://www.ctvnews.ca/politics/article/canadas-emissions-reduction-plan-falling-short-environment-commissioner/.

of climate change, has not been disbursed or translated into visible, constructive projects that can be visited and felt on the ground. This gap between promise and action undermines trust in the COP process.

The stakes are too high to give up. Climate change continues to accelerate, and the window for action is closing. The need to double down on efforts is clear, and there are moments of progress that give us hope. At COP28 in Dubai, I felt proud that Colombia assumed the presidency of the Beyond Oil and Gas Alliance (BOGA), while my province of Quebec stepped into the role of vice presidency. We must continue.

Moving forward, COPs can become more effective by focusing on greater transparency in tracking financial commitments and holding countries accountable to their NDCs. This also means limiting the influence of industries that have a vested interest in maintaining the status quo, such as fossil fuel companies, and thinking critically about whether fossil fuel lobbyists should be allowed at COPs. A positive future for the COP process would involve turning these summits into platforms that not only raise climate ambitions but also deliver tangible results—taking real steps towards decarbonization, moving towards climate justice, and a fair transition to a more sustainable, equitable world. If we focus on these priorities, COPs can still evolve into the engines of real climate solutions that the world so desperately needs.

4.7. Conclusion

As I reflect on my journey through the COPs, from my first experience in Lima in 2014 to COP29 in Baku, I am reminded of the importance of international cooperation on existential issues affecting us all. The annual COPs play a critical role in the global effort to address climate change. These gatherings bring together world leaders, scientists, parliamentarians, activists, and civil society to negotiate, collaborate, and commit to actions that aim to mitigate the devastating impacts of climate change. COPs have become the cornerstone of international climate diplomacy, providing the platform where crucial agreements are forged, and where countries are held accountable for their climate pledges.

Over the years, I have witnessed both the immense potential and the profound challenges that these conferences represent. From Lima to Paris, Madrid, Glasgow, Sharm el-Sheikh, Dubai, and Baku, the progress made in these global forums—despite the challenges—is essential for maintaining the momentum needed to keep global warming within manageable limits. These conferences are now giving more space to vulnerable nations to voice their concerns and advocate for climate justice, ensuring that the impacts of climate change on communities, especially in the Global South, are at the forefront of the negotiations. Additionally, the growing engagement with civil society, Indigenous Peoples, and youth movements has made COPs more inclusive, and increased the push for more ambitious climate action.

However, the COP process is not without its challenges. While countries have made bold pledges, the implementation of these commitments often lags behind. The growing presence of fossil fuel lobbyists at COPs and the struggle to secure adequate financing for vulnerable nations remind us of the obstacles we face in achieving true climate justice. Despite these setbacks, I remain hopeful. The global community continues to raise ambition, and the conversations at COPs grow more inclusive, with an increasing focus on the voices of those most affected by climate change.

COPs are essential in keeping the international community accountable and pushing for the actions necessary to limit global warming, protect biodiversity, and ensure a just transition to renewable energy.

For me, COPs represent a space where my lifelong dream of a cleaner, healthier world comes closer to becoming a reality. Growing up in Peru, I developed from a young age an interest in the environment and how to protect it. It is here that I first imagined a future where humanity and nature coexist in harmony. This dream, shaped by my upbringing and my career in environmental engineering, continues to fuel my work at these conferences.

In the end, the dream of a cleaner world is not just mine—it belongs to all of us. It is a dream rooted in the understanding that the health of our planet is inseparable from the well-being of its people. As a parliamentarian, an engineer, an advocate for the environment and a grandmother, I will continue to work towards this vision, ensuring that the actions we take today secure a vibrant, sustainable future for generations to come.

CHAPTER 5

COPs Are Important Gatherings: They Keep Climate Change on the Map

Elizabeth May

Media commentaries misunderstand the process. Each Conference of the Parties (COP) has its own dynamic. Media often blame the most convenient target, but the actual dynamic of a COP depends so hugely on the character and the talents of the president of the COP. The United Nations (UN) and multilateralism are criticized but multilateralism is needed to move forward. The consensus system is a big plus, as it keeps large economic powers like the United States in check. COPs are important gatherings. As 190 countries cannot negotiate virtually, COPS are in-person events, which I think need to be understood as a form of governance over climate actions. The treaty is critical, and we are not living up to our treaty commitments. But the treaties are not just paper documents, they have created a living, breathing governance structure, like a congress or a parliament. That means that once a year, a global climate parliament convenes and we are part of it. A widespread criticism is that we have had enough of these, they are not working. However, what is the alternative? What would it be? There are 190 countries all willing to come together in the same room, and we have an established process that was not easy to build. Instead of dismantling it, we should be improving it; to do that, we need to look at the glitches.

I was involved before the United Nations Framework Convention on Climate Change (UNFCCC) was finalized at the 1992 Earth Summit. I first attended the COPs as the executive director of the Sierra Club of Canada. I attended COP11 within the Climate Action Network. My job was to keep an eye on the negotiations. Other climate activists felt I would be the best one to keep an eye on the deputy minister of Environment Canada and monitor progress. COP11 was hosted in Canada. We did not worry about the president, Stéphane Dion. Even though he is not known as a good communicator, he was resolute, is not inconsistent, and has great integrity. We were worried about Deputy Minister Sammy Watson. NGO observers usually do not have access to negotiating rooms, however, COP11 was more accessible. I was pretty much getting into every room because that is how the Canadian government treated NGOs at the time.

I have known Bill Clinton since I was 17 when my mother worked for him in the campaign for George McGovern, and I invited him as former president. At first, he turned down the invitation, but then he suddenly changed his mind. The Prime Minister's Office called me and said that Clinton had an invitation from me, he wanted to come, and they really wanted him to be there. They said, "Can you help us get him there?" That was weird because it meant that I had to organize private planes. I had to call the Sierra Club of Canada's biggest donor and ask for his credit card number over the phone so I could get a $20,000 deposit on a private plane to get Bill Clinton into the room. So I was an NGO observer keeping an eye on Sammy Watson while trying to get Bill Clinton in the room. It was eclectic.

Since then, I have participated in several COPs. When I went to COP15 in Copenhagen, I was not a member of Parliament (MP) yet. I was there with the Global Greens with an observer pass. My role was to keep an eye on the Canadian delegation led by the minister of the environment, Jim Prentice. It is important to keep an eye on the delegations during the negotiations to be able to track and share information with other climate activists and keep the pressure on the government. There are several moving parts during these conferences; it is important to know what the country is always doing and pay close attention to what we hear in the corridors.

Under the Harper government, MPs were not allowed on the delegation. Since becoming an MP, which is a longer story, I had to be invited on other countries' delegations or under NGOs as an observer to participate in the negotiations. In Durban, I went on the delegation of Papua New Guinea. I basically did diplomatic couch surfing. In Warsaw, I went with the Global Green Party, but once I got there, the delegation of the Government of Afghanistan approached me and asked me if I wanted to join their delegation. They only had three people and only one of them spoke English well. Even if the UN has eight official languages, everyone speaks English. It is really important for everyone in the room to understand legally binding obligations before accepting them.

The International Forest Convention fell off the rails in 1990, but when the World Trade Organization (WTO) was created, it exercised its clout. It told trade ministers that enforcement provisions that were allowed in the Montreal Protocol to save the ozone layer could not be used for Kyoto. The lack of enforcement mechanisms completely undermined the system. If we want the climate agreement to be more ambitious, we need to redress the influence of the WTO. In reality, the WTO is not allowing provisions to protect the environment. The WTO remains, even today, a vector to advance fossil fuels. As long as the WTO is allowed to sabotage climate action, we are going to have a hard time achieving the objectives.

The WTO's 13th Ministerial Conference took place in Abu Dhabi, in the United Arab Emirates. Since its creation, the WTO has always used trade rules to sabotage climate negotiation. It has blocked any effort to put meaningful sanctions or penalties on countries. The 13th Ministerial Conference would have been a perfect time to ensure that trade negotiations and trade decisions become climate compliant. It would require the WTO to recognize that the Paris Agreement had precedence.

We are lacking any enforcement mechanism built in the UNFCCC. We end up with toothless agreements and commitments that countries can simply fail to achieve and get away unscathed, basically because the public does not know. However, the Canadian public gets a sense that we are not doing well; they kept seeing emissions going up. Because there are no penalties, Canada, like every other country in the world, is free to ignore its commitments. We have long been the only industrialized country that initially ratified Kyoto and has continued to see emissions rising, when other countries, for instance European countries, have been able to bring them down. It is simple, we did not have any penalties in the other protocols, and were, along with any other country, free to ignore our commitments even though they are legally binding.

COP11 in Montréal (2005) was Canada's only time hosting a climate conference. It was a great success, and parties set an ambitious timeline. Stéphane Dion was a good chair, and he did not put us off the rails even though COP11 took place in a complex Canadian political context. Prime Minister Paul Martin's minority government fell on the opening day of COP11. It was a different kind of COP. The United States' opponents to climate action were derailed by Bill Clinton, who gave a strong speech supporting strong goals. However, the Russian delegation tried to block things for a long period of time. The Montreal conference ended with an ambitious timeline to replace Kyoto within four years.

COP15 in Copenhagen was also a turning point. We saw there the convergence of several forces determined to undermine the whole climate negotiation process.

The fossil fuel forces were really coming back; they really wanted to make sure that the kind of success we had in Kyoto in 1997 was not going to happen again, and, relatively speaking, they also wanted to make sure that the success seen in Montreal in 2005 could not happen again either. The fossil fuel actors lobbied in the lead up to COP15. The fossil fuel lobby hacked into the computers of scientists. They tried to hack the computers of Dr. Andrew Weaver and the Hedley Centre, and they attempted to discredit the entire climate scientific community. They discredited brilliant scientists, attacked their motives and distorted the facts. They broadcast it widely as a "leak," as opposed to a hack. There was no sign whatsoever that the scientists who were communicating with each other in emails were anything but honest, ethical and hard-working. I read all the emails—there were about 4,000 emails that were released—and the efforts to discredit focused only on one or two lines, pulled out of context. The fossil fuel lobby was looking at some distortions in paleo-climatology, looking at several complex variables and estimates from centuries ago. Scientists were basically dealing with many, many, many complex and different data points. Nobody was distorting anything, but the attack on the scientists was brutal, and it was right before COP15 started in Copenhagen.

Copenhagen also occurred under the auspice of a new Danish government. It is important to highlight that, at the end of every COP, we do not necessarily know who is hosting the next conference. By 2007, we learned that Denmark offered to host. At that point, they had a left-wing government in power, and just to confuse things, the prime minister's name was Rasmussen. However, between the announcement and the opening of the COP15 in Copenhagen in 2009, the government was led by a different prime minister with the same family name, Lars Løkke Rasmussen. Unexpectedly, the Danish government did a lot of things to undermine the negotiations.

Prime Minister Rasmussen fired the minister responsible for the COP and took over the proceedings. He imposed martial law to crack down on social movements; several delegations were not let inside the building, keeping people outside freezing, like the Chinese and Indian delegations. I came as a Canadian, fully prepared for cold weather, but they kept us locked outside the building routinely five hours every morning, even if we had credentials. So, in my view, Copenhagen's failed negotiations were definitely the fault of the Danish government.

COP15 was an absolute failure, even from a Canadian perspective. When I was in Copenhagen, I was part of the movement Global Greens. At the time, former environment minister for the Harper cabinet, Jim Prentice, was the head of the delegation. He brought Mike Holmes of *Holmes on Homes* as a "climate expert," who claimed sunspots cause global warming. It is important to understand that the scope of the negotiations is set ahead. The extent to which the Harper government was able to sabotage the negotiation was never really understood or well known because it was very subtle. Harper's team did not stand up and give a big thundering speech. They fanned out through many rooms and engaged in low diplomatic tones. I witnessed how Canada sabotaged the negotiation in every room. Stephen Harper designated Michael Martin as climate ambassador. He was tapped by Harper from Global Affairs and was mandated to sabotage Copenhagen. In 2014, Martin was appointed deputy minister of Environment Canada. There was a stark difference between COP11 and COP15. I would say that at the end of the week, on Day 10, in relation to one piece of text, Canada, as a negotiating party, added square brackets around paragraphs 1 through 4. It does not sound like much, but it completely derails the negotiation process. It means that if we do not have an agreement on any of that text, we have to start over. Negotiators from Canada started placing square brackets around a lot of text, after having previously been silent for days. Canada sabotaging things is something I find quite interesting. Canada still has an outsized role in the world of negotiations. We are famous for punching above our weight, we punch above our weight when we try to derail something. It affects Canada globally. The Harper years were very effective and devastating at slowing down climate action globally. It was never really understood why the Harper government was able to sabotage COP15 as well, but Canada's square bracket game completely derailed the negotiations.

The next COP, COP16, was in Mexico. If anyone had asked me in Montreal what country was going to be more supportive of getting good negotiations—a strong protocol to follow up on Kyoto as Kyoto wound down—Mexico or Denmark, I would have said Denmark every time. Denmark killed it, but Mexico saved it. Particularly Patricia Espinosa Cantellano, Mexico's foreign minister, who went on to be the secretary general for the UNFCC from 2016 to 2022. She ran the negotiations as the foreign affairs minister in Cancún and managed through force of will and tireless work to get things back on track. Her work led us to Durban, which was also successful despite the relentless sabotage by Canada and the United States. The Government of Canada pressured India not to agree to anything, which enabled the Harper government to say that the Government of Canada was not going to support any progress in the negotiations until China and India moved while they were busy trying to encourage China and India not to move. Can you even imagine the tactics? Durban succeeded again because of a very strong presidency and a very supportive host country. Sadly, things became worse in Poland. Once again, another president tried to sabotage the negotiations. The Polish presidency did not want the negotiation process to succeed because of wedded interests in coal. I am delighted that the negotiations at COP28 in Dubai went as well as they did, because it sure did not look like we were going to a country where a government wanted us to succeed, but the Global Stocktake was an important step.

During the Bonn Climate Change Conference, Canada was still president of COP. Then Environment Minister Rona Ambrose, as president of the COP, presided over the conference. In Bonn, Canada changed its target, which was weird because our targets under Kyoto were baked into the treaty. But when Rona Ambrose went to COP and said, we are no longer bound by our Kyoto target without a vote in Parliament, it shocked everyone. The Harper government replaced the Tokyo target of 6% under the 1990 levels by 2012 with the new target of 20% by 2020 below 2006. After Kyoto, everyone used the same base year. So Canada alone sabotaged the entire framework. It took a lot of countries quite a while to realize that Ambrose said a 20% reduction by 2020 below 2006, without any penalty for doing so. The United States followed suit. However, when we were in Copenhagen, President Obama changed the year again and used the 2005 levels, so Canada ended up using the same yearly target. It is important to compare the GHG reduction target to all European countries who were achieving the 1990 targets. There is a lack of understanding why it matters that every country sticks to the same year. By using the same year, it is easier to compare targets. I translated the target back to 1990, and it is very clear how low Canada's target was: Our emissions would even be higher than 1990 levels. Virtually every other country in Annex I was on its way to achieve a reduction below 1990. It is important to point out that China is not in Annex I. However, Canada was way above the 1990 target. The other country way above the target was the United States, but they never ratified Kyoto. We signed it, we ratified it, and then we withdrew after illegally and unilaterally changing the target without any consequences. It is quite unbelievable how much damage Canada has done.

COP21, the Paris Agreement, was a huge success because it made the goal to hold the temperature below 2°C possible. However, the one weakness was the architecture of the targets. In Paris, The Harper government decided to give some flexibility to countries. Every country must file their own target; unfortunately, a global target was not baked into the agreement. I think it was a significant mistake. COP21 took place just a few weeks after the new government was sworn in. When Justin Trudeau was elected, there were hopes that there would have been a change in direction, but Trudeau kept the same target in place as the previous government. In addition, the Trudeau government retained Michael Martin as deputy minister of the environment. However, what the Liberals put in their platform was so weak that it should not have fooled any environmental group. Before going to Paris, they convened a federal–provincial meeting. Michael Martin, as deputy minister, was the lead. He held the pen on the Pan-Canadian Framework on Climate Change, and he designed the whole process with every province and territory. Minister Catherine McKenna wanted to get rid of him. When I talked to Jerry Butts and Justin Trudeau about what I perceived to be a problematic army of marching zombies still taking Harper's order, they basically told me that they would not interfere with the civil service. They promised during the 2015 election that they were going to respect the independence of the professional civil service. They said that they would leave it to Michael Warnock, who was the clerk of the Privy Council of Canada at the time. Michael Martin was not just a problem in the civil service, he was a saboteur. He sabotaged Copenhagen and he was trying to sabotage Paris. There were some good people on Catherine McKenna's personal staff. McKenna stepped out ahead of her bureaucracy with the encouragement of one woman on her staff, named Jane McDonald. At that point, Canada managed to come back. They did not change the Conservative target, which did not align with the Paris Agreement, but, I think, Canada was the first industrialized country to agree to add the 1.5°C target in the declaration. That was not what the bureaucracy at Environment Canada wanted. In fact, the bureaucracy tried to backpedal from that, but I had already texted a *Globe and Mail* reporter and a CDC reporter who were in Paris to say that Canada had just taken that position. It was already in the news when Environment Canada tried to deny the commitment.

It is hard to keep track of Canada's target. At first, they set 40%, then they filed 40%–45% again, but note that we are still using the 2005 baseline year, and our target remains weaker than the one of the United States (50%–52% below 2005 by 2030). These were the same nationally determined contributions that were put in place by Leona Aglukkaq. It was the fourth time; I think that Harper had weakened Canada's target. In reality, the manipulation of Canada's GHG target continued under Prime Minister Trudeau. Citizens cannot keep track of GHG targets. They are really boring for people who cannot spend their time assessing the progress made by governments in Canada. But there should be a simple metric. What did we promise? Are we keeping our word? Are we hitting the target? These are the wrong questions. The media falls for that every time. They focus on whether we are on track to hit the target. One, no we are not. Two, the government picked a target that is inconsistent with what we agreed to in Paris. We agreed to keep the temperature down to 1.5°C and to not exceed 2°C. Full credit to the French government and the brilliance of the chair, Laurent Fabius, who managed to keep all those countries and their delegations on track. The temperature targets in Paris are the targets that matter. By signing the Paris Agreement, Canada committed to keeping the global average temperature below 1.5°C, at most 2°C. This commitment required a carbon budget. And yet, this government was able to get away with bringing in a phony climate accountability legislation, which has no carbon budgeting. It also did not include the temperature target that is so critical to the Paris Agreement. We only talk about net zero by 2050, while continuously not achieving our commitments.

The ability of our own government to repeatedly commit massive fraud is a combination of lacking enforcement mechanisms in the treaty and having had shifting goals that are hard to communicate. The complexity of these shifts has easily fooled the national media. On top of that, relativism is omnipresent. Everyone said, "Liberals are not doing enough to make the Green Party happy, but in reality they should be doing so much better." I knew this was right just after we started to realize that Trudeau was not going to keep his climate commitments. Basically, the government promised much but delivered little. It said strong things about the need to take climate action, but it did not actually do anything to be a climate leader and put in place the actions that are actually needed to fight climate change. Canada still should win. If this government, even after all the betrayals and building the Trans Mountain pipeline and everything else, if it was to stand up and say,

> [Y]ou know what, our target isn't good enough. It's inconsistent with the Paris Agreement, we looked at carbon budgeting. We know we have to do more, so we're going to shut down the Trans Mountain pipeline. It's a mistake to build it if we're not going to use it [...] and we're going to ban fracking from coast to coast, because we've determined that this technology is inherently unsafe for groundwater supplies, inherently unsafe for the atmosphere, and by 2030 we're going to stop producing bitumen from the offsets [...]

That would be the right thing to do. There are steps we can still take, but it seems to me that most of the media, even sometimes independent media, such as the *National Observer* or the *Tyee*, get misled by the question, "Are we on track to hit our target?" They look at the target in the NDCs report tabled in 2020, and that is what they report on. But Environment Canada does not tell the truth anymore; there was a time when Environment Canada told the truth. But Environment Canada, as an institution, does not. The department is perfectly happy to say, "We're on track," because it is happy to say that we are going to achieve the net-zero target by 2050. However, it is worth mentioning that it is completely irrelevant if we do not cut our emissions by at least 60% below 2005 by 2030, and if we want to have any hope of doing our minimal fair share to hold the global average temperature at 1.5°C or below 2°C. Instead, our current target, which ranges between 40%–45% was changed because John Kerry and the US administration threatened the Prime Minister's Office and did not invite Canada to speak at the Earth Day Summit in 2020. We are so far away from the United States' commitment, which, before the Trump administration, was 50% below 2005 levels by 2030. So the most we got out of the PMO was 40% to 45%. The United States looked at it and said 45% is okay; it is close enough to 50%. Sadly, this is the last time that Canadians will ever hear anyone in Canada talk about 45%, yet they say we are going to hit our target. Realistically, we are not on track to hit the 40% target, but even if we were, this target remains inconsistent with the commitments we made in Paris.

Social movements are an important part of climate conferences. It matters a lot to have good turnout at demonstrations. It mattered in Glasgow, and of course, when conferences happen in countries where demonstrations are not allowed, it changes the dynamic. Negotiations go better in general when we are in a country that is democratic, that allows for people to be on the street. Canada is a good place for public mobilization, because we need to achieve the targets of the Paris Agreement, and we need to put pressure on the governments. Sadly, the exception was in Copenhagen, which was so weird because you would think that Denmark would protect civil and democratic liberties, but they did not do so during the conference. They suspended civil liberties, banned demonstrations, and tear gassed people. They cracked down on protests and civil disobedience, even before the negotiations began. So, when we think about an initiation to COP and how the public understands what COPs are, one thing that is helpful to understand and to make very clear is that social protests are part of the process.

It is, unfortunately, an in-person event. The other thing that I think we may overestimate, even overemphasize, is the number of fossil fuel lobbyists who attend these conferences. It gives the public the impression that fossil fuel lobbyists are influencing the negotiations but they are not. They are in the city, they are not in the room. They are twisting arms where they can. It is not great, but they do not necessarily achieve their objectives.

Some COPs are more important than others, and that is another key element. For me, there is a decision point: Is there a reason to go there? Why am I going? Is there a reason for going? I decided not to go to Azerbaijan because I do not want to attend a conference in a country that just so recently committed genocide against Armenians. Russia vetoed every other European country. It would help if Canada supported an eastern European country to host a COP. Technically, it would take five years for a European country to host again. The rotation to host COPs is based on geography. So even though it does not look like they moved very far from Sharm el-Sheik to Dubai, those were, in UN terms, different regions. The United States has never hosted, and Canada has only hosted once. COP should come back to North America and Canada should definitely be the next host. If it is held in Canada, citizens can demonstrate and mobilize, like the mobilization in 2020. We saw an amazing mobilization all around the world, and Canada is a good place for it.

So where are we in all of this? We are off track. How do we communicate what the COP process is and why COPs matter? Well, it is important to highlight that multilateralism matters. Why does the UNFCCC matter? How do we make the Framework Convention on Climate Change have teeth, when it was denied teeth in its embryonic negotiating process and its enforcement provisions? The UNFCCC, the Kyoto Protocol, and the Paris Agreement are seen as the dominant international treaties, and they are international law. They are as long as we do not let the WTO sabotage climate action, otherwise we are going to have a hard time achieving the promises under Paris and simply understand how to get there.

The presidency of the COP is so important that I would love to see the climate conference stop moving to different locations. Let us have all the COP negotiations in the UN General Assembly buildings in New York or in Bonn. COP22 was held in Bonn because the host country, Fiji, did not have the hotel space and negotiation facilities to host the conference on the island. There should be more or less permanent locations for COPs but we should absolutely have different countries host. We would not be stuck in difficult situations where thousands of delegates are dependent on the host country. COPs are also susceptible to criticism, such as that it is a jet setting waste of time and resources in moving people around to different countries.

The other thing, of course, is what I would have never foreseen, even as recently as Paris in 2015, which is a Trump administration, not only attacking climate action but also the WTO. I really dislike the WTO, but if you are going to have a global system of rules, it is incredible that you suddenly have the United States against any form of multilateralism, including trade rules that they wrote to advance their own interests. I find myself in a funny position to say, "yes, let's protect multilateralism and the multilateral global acceptance of agreed upon rules." These rules, like in 1987, saved the ozone layer; these rules are handy. We need a post-pandemic discussion. We need to rearrange the economic rules of the world. We cannot rearrange climate change, we cannot rearrange physics and chemistry, but we can rearrange the economic rules, the financial flows, we can tax international shipping and international aviation, we can give more money to developing countries for adaptation and help them move toward renewable energies to mitigate climate change, but we cannot rearrange the climate. Canada must and should play a significant role moving forward. We should host another climate COP.

CHAPTER 6

What It Takes to Represent Canada at Climate COPs: A Federal Government Perspective

Catherine Stewart

Since the Paris Agreement was adopted in 2015, I have attended every United Nations Framework Convention on Climate Change Conference of the Parties (known as the Climate COP), serving in various senior executive roles for the Government of Canada. From my observation, they have evolved significantly over the past eight COPs to be more complex, involving more people, interests, and interlinked issues. Not surprisingly, they have also come to serve many different purposes. For some, COPs are like tradeshows to promote new and innovative ideas, technologies, research, and policies and to network with a broad range of people, such as innovators, investors and foreign governments. For others, they are an opportunity to draw global attention to the widening scope of the impacts that climate change is having on all aspects of human life. Above all, however, COPs have become recognized as the largest and most important annual gathering on climate change on the planet, where countries come together to demonstrate their commitment to address climate change, defend their climate interests, and put pressure on each other to do more. They have also become a venue for civil society to actively hold leaders, ministers, officials, and other actors to account and advocate for stronger commitments, transparency, and accountability.

For governments (or parties), this hive of activity happens within the negotiations space, where officials and ministers come together to agree on the next priorities and actions that countries will take collectively to tackle the global climate crisis. They come face to face with counterparts to express their respective positions, engage in frank dialogue, and attempt to find compromises. As a party-driven process, the level of access to these spaces can vary. In some instances, the meetings are restricted to party heads of delegation only. Whereas in other cases, the meetings can take place in large plenary rooms, which seat hundreds of people. But conversations among parties can also happen more organically. For example, the COP presidency can purposefully call a plenary session with all parties and hold off starting the meeting so that heads of delegation can speak with each other to try to hash out an issue and come to consensus. You will often see small huddles happening in various corners of a plenary, where intense conversations are happening with a view to finding resolution. Valuable discussions among governments also take place in the corridors outside of meeting rooms as delegates wait to be allowed in. Governments can also schedule bilateral meetings with counterparts in a dedicated delegation space in order to share positions and try to form common alliances and strategies to push forward a common agenda.[1] But conversations between governments also take place during high-level side events and organized discussions. Countries host pavilions, for example, which provide opportunities for

---

1. Many governments and organizations pay for dedicated delegation spaces at COP. For Canada, the delegation room is a space where negotiators can focus on their work, have lunch, or take a breather before their next series of negotiations. It is also an area where ministers and heads of delegation can host a meeting with a counterpart, otherwise known as a "bilateral meeting" or a "bilat."

panel discussions, "fireside chats" (one-on-one conversations in front of an audience) on specific themes and policy priorities. These side events can also bring in representatives from other parts of society, providing greater depth and richness to the discussions.

Each COP has its own trademark and flavour, depending on the geopolitical context and the approach and priorities of the host country holding the COP presidency. But no matter the year-over-year differences, the effort and energy required to ensure that Canada is well represented and influential in this space can be significant. What does Canada's public service do to support a large meeting like COP? What are some of the decisions that need to be made and considerations before heading to COP? What is it like to be on the ground at COP as a negotiator or to support a minister? What happens in the aftermath? How do we keep the momentum going all year round until the next big COP?

6.1. Let Us Begin with Formal COP Negotiations

COP negotiations have represented the biggest opportunity to press for more collective action on climate change. There are many different elements that are currently under negotiation based on commitments in the United Nations Framework Convention on Climate Change (UNFCCC) and the Paris Agreement, but there is usually one prominent negotiated text every year, which is sometimes called the COP outcome document, or the final decision document negotiated by the COP presidency. As a negotiator, there is a real sense of accomplishment when the gavel comes down to mark the end of COP and the final text is adopted. It represents hours and hours of painstaking negotiations and sleepless nights over concepts and initiatives and even over words and syntax—indeed every comment and every word matters. It represents months of discussions among parties to understand each other's perspectives and "redlines"—the limits of what they can or will politically accept. For Canada, that final document is also the culmination of whole-of-government coordination and collaboration and consultations with subnational governments, stakeholders, and Indigenous people. Most of all, the final document signifies our intentions and commitments as a global community and sets out the ambitions and standards that we all want to live by. We often say that you will know the sign of a good negotiation outcome if "everyone leaves *equally* unhappy." Compromise is at the heart of any final deal and COP negotiations are no exception.

The outcome at COP28 provides an interesting example of the art of compromise. In Dubai, parties gathered to take stock of where we are as a global community in implementing commitments to address climate change. As mandated by the Paris Agreement, this "Global Stocktake"[2] was also an opportunity to identify solutions to address the gaps in ambition and inform climate policies and plans that were due in 2025.[3] One of the more notable outcomes in the Global Stocktake includes reference to "transitioning away from fossil fuels in energy systems, in a just, orderly and equitable manner, accelerating action in this critical decade, so as to achieve net zero emissions by 2050, in keeping with the science." You could spend some time dissecting this sentence to understand why every word is important. In one instance, it is in line with the International Energy Agency's Net Zero Roadmap,

---

2. According to Article 14 of the Paris Agreement, the Global Stocktake is designed "to assess the collective progress towards achieving the purpose of [the Paris] Agreement and its long-term goals. Those goals include cutting greenhouse gas emissions to limit global temperature rise to well below 2 degrees C and ideally 1.5 degrees C; building resilience to climate impacts; and aligning financial support with the scale and scope needed to tackle the climate crisis."

3. Also referred to as "nationally determined contributions" (NDCs). Under the Paris Agreement, all parties shall put forward an NDC. A country's NDC is a self-defined climate pledge that details what that country will do to reduce its emissions and adapt to the impacts of climate change.

which implies that there will continue to be a demand for oil and gas even in a net-zero 2050 scenario (albeit at a much-reduced level).[4] In another instance, it also considers the Sixth Assessment Report of the Intergovernmental Panel on Climate Change (IPCC), which references the impact of abatement technologies in the oil and gas sector on global warming projections. This formulation cleverly steps away from the fossil fuel "phase-out" language that many were looking for, but which was equally unacceptable for many others. It nevertheless underscores the central role of fossil fuels in climate change and the collective intention to move away from them. As many have noted, this text is unprecedented, marking the first time that all parties to the UNFCCC have come together to indicate the need to address fossil fuels and to signal intentions for the future.

As a government representative, I operated under a negotiating mandate, which was approved by Cabinet and my activities and the actions of my team were guided by the priorities of the minister of Environment and Climate Change Canada. Before we set out to engage in negotiations, my team would prepare a "Memorandum to Cabinet"—a detailed document that lays out our proposed objectives, weighs the options, and recommends a way forward. The minister of Environment and Climate Change Canada would ultimately bring a final draft of this document forward to seek support from their cabinet colleagues. With this direction from Cabinet in hand, my team would then draft instructions outlining the space in which negotiators would operate in, what they can push on, what they cannot accept, and when to draw a red line or seek further instructions from the political level. These instructions are useful for the cohesion of the team given that each negotiator will be focusing on specific issues (such as Article 6 of the Paris Agreement on carbon markets,

---

4. IEA, *Net-Zero Roadmap: A Global Pathway to Keep the 1.5 °C Goal in Reach, 2023 Update* (IEA, 2023).

or the information that countries would have to periodically provide in reports on their actions and progress) and there can be crosswalks between negotiation items and sometimes trade-offs as you get closer to the end of negotiations—if you are willing to give on one side, you can gain on the other.[5]

For anyone who has ever spent time in the negotiation trenches, you will know that there is a great deal of coordination that happens among negotiators—clever tactics and strategies are used to get better outcomes, intelligence is shared regarding the trends and agendas they are seeing, alliances and networks are built around common issues. There is nothing more gratifying than seeing the newer, younger negotiators on your team develop into collaborative but strategic diplomats over a two-week span, doggedly trying to bring opposing parties together to find compromise all under the steady guiding hand of the more seasoned negotiators. During those early morning hours, with very few hours of sleep in several days, there can also be a healthy dose of memes that get passed around and other creative attempts at humour!

From the outside, it is hard to imagine how you can get an agreement among 196 parties. The existence of negotiating blocs helps to corral like-minded countries together to develop common positions and approaches. These groupings also help to create efficiencies, as the leader of the bloc can be empowered to speak on behalf of its members. One of the biggest blocs is "the Group of 77 and China," which has grown since its inception to include over 130 developing nations. Within that group, you will find smaller groupings, including the Alliance of Small Island States and the Africa Group. If you think a two-week COP is long, many negotiators in

---

5. Article 6 of the Paris Agreement promotes cooperative approaches between countries to help drive down emissions reductions. Progress reporting under the UNFCCC and Paris Agreement comes in many forms. For example, Canada submits a National Inventory Report to the UNFCCC annually to report on Canada's greenhouse gas emissions.

the G77 and China spend one week prior to COP pulling together common negotiating positions and strategies. That is a long time to be away from home! For our part, Canada works within the "Umbrella Group," which is a loosely knit group of countries that are not members of the EU or the G77 and China.[6] Canada's federal negotiators share information and ideas year-round with the Umbrella Group and daily during COPs. While the Umbrella Group shares similar views on several issues, and issues common statements on occasion, Umbrella Group members retain their individual positions and will speak freely on their own during meetings as required.

As a woman in this space, I can unequivocally say that there is a lot more work that needs to be done to increase female representation and support across the board. I know there are some women negotiators who are not comfortable negotiating into the long hours of the night and then having to find their way back to their hotel rooms alone. I have also witnessed deliberate intimidation tactics and men speaking to only the other men in the room. While there has been some progress during my time, such as greater sensitivity to having more inclusive and diverse panels, overall progress has frankly been very slow. There are more advocacy groups, including SHE Changes Climate and Women Leading on Climate, which help bring women together to better support each other. I have also found that even the smallest of gestures can help to empower women such as referring to a woman's statement when you are speaking in plenary and passing along a positive WhatsApp message to a colleague after she has spoken. But clearly more needs to be done on this front. For our part, Canada has more recently had a good gender balance on its federal delegation, which is a reflection

---

6. The Umbrella Group membership includes Canada, the United States of America, Japan, Australia, New Zealand, Ukraine, the United Kingdom, Kazakhstan, Norway, Israel, and Iceland.

of Canada's federal public service writ large—there is nothing more powerful than showing by doing.[7] At the practical level, we consider various ways to ensure the safety of team members. For example, we have put in place a "buddy system," including with other parties to ensure that negotiators are not left alone to return back to the hotel late at night, once the negotiation session is over.

Furthermore, since 2015, Canada has consistently fielded a diverse and inclusive delegation with representatives from across society—parliamentarians, senators, subnational governments, leaders of national Indigenous organizations (NIOs), business representatives, labour groups, youth, civil society, and other stakeholders. The composition of the delegation is a ministerial decision, and the public service helps to secure and implement that decision. Each year, more and more people are interested in attending COPs—something I attribute, at least in part, to the increasingly visible and worsening impacts of climate change. The number of people on Canada's delegation has ranged from about 200 to 400 people at each COP and the delegation lists are published on the UNFCCC's website once the meeting is over. Part of my job as chief negotiator was to ensure good outreach throughout the year with all orders of government, NIOs, and key stakeholders to help inform advice to ministers on Canada's positions at COP but also to prepare the delegates ahead of COP on Canada's priorities and objectives. At COP, delegation management includes frequent briefings on the status of negotiations and activities and ensuring that everyone wearing a Canada badge is aware of the rules of the game, including the role of the head of delegation.

---

7. Canada's climate finance has also supported training of women negotiators from developing countries to provide them with the skills they need to influence international climate discussions. This support has been in place over successive years and has also enabled some participants to travel to and attend COP.

Returning to the specifics of negotiations, after dedicated intersessional negotiations and workshops throughout the year, the preparation in the few months before COP becomes more intense. Ministers who intend to go to COP to head the national delegations will likely want to be involved in negotiating the final outcome at COP and will want to be prepared to engage. Skilled public servants prepare briefing materials to help capture the state of play of negotiations for ministers based on their priorities and interests as well as their experience with multilateral negotiations.

The COP presidency will also use different strategies to ensure ministerial engagement. They will generally host an in-person[8] pre-COP one month before the main meeting to ensure broad understanding of the issues at hand and the differences among parties and start identifying common ground ahead of COP. The COP presidency will often ask pairs of ministers (from a developed and a developing country) to engage further with countries and come up with solutions to divisive issues. In those final hours of negotiations at the COP itself, I have also seen a COP presidency hold a ministerial meeting at 1:00 a.m., crowd ministers in a small, hot, uncomfortable room to entice them to show flexibility, and move the needle and secure breakthroughs in negotiations. Since 2015, COP presidencies have also nominated high-level champions to help connect the COP to broader groups outside of government. They help by translating the presidency's priorities into specific discussions, themes, and actions that resonate with these groups

---

8. In a multilateral setting, the bulk of the meetings are in person, which helps to "level the playing field" among all parties. In-person meetings bring everyone into the same time zone and avoid spotty Wi-Fi access issues that some parties experience. Some developing countries are also provided with funding to attend COPs through UNFCCC's Participation Fund, recognizing the value of in-person meetings for relationship building, bridging divides, and increasing ownership to find consensus.

and better connecting activities outside of negotiations with political decision-making.

6.2. Moving Beyond Formal COP Negotiations

Before and during COP, there are other initiatives that are brought forward to help to raise global ambition, including the introduction of declarations and statements. These initiatives take on a life of their own: They can be advanced by the COP presidency or any group of actors or countries on any given topic that touches climate change. Since the adoption of the Paris Agreement, Canada has introduced a good number of initiatives to accelerate global action and results, which have been launched at the political level. For example, in 2017, Canada and the United Kingdom joined together to introduce the Powering Past Coal Alliance (PPCA) to promote the phase-out of coal energy. Sparked during a ministerial bilateral meeting in 2017, this Alliance now has over 180 members committed to global coal phase-out. Canada also worked alongside the European Union and China and launched the Ministerial on Climate Action, which brings together ministers from developed and developing countries to discuss COP negotiations and push for ambitious efforts at the global level. In 2021, Prime Minister Trudeau introduced the Global Carbon Pricing Challenge to promote carbon pricing as a valuable policy tool to reduce global emissions and to promote innovation and affordability. Canada is also the co-chair of the Concrete and Cement Breakthrough, launched at COP28 in 2023, which aims to support the decarbonization of this important industry. These initiatives, declarations, and statements can serve to raise the level of ambition in the COP context. For instance, over the past seven years, the PPCA has served to raise global visibility and momentum around the importance of coal phase-out. So much so that the COP final decision at COP26 in the United Kingdom acknowledged the need for coal phase-down. The fact that the COP outcome document mentioned coal marked an important step forward, as unabated coal power still represents the largest source of CO_2 emissions globally. By representing aspirations for a cleaner future, these activities outside of the COP process can have an influence on COP outcomes.

Just as Canada advances initiatives, so too do other countries and stakeholders. We also receive more and more declarations, statements, and initiatives for Canada to consider joining or co-sponsoring; each one needs to be triaged and consulted to determine if it aligns with Canadian priorities.

COPs also serve as an annual moment for countries to bring forward announcements, whether it be enhanced support for developing countries or the launch of a new policy initiative and there is a lot of work required across government to get us to that point.

6.3. Then There Is the Programming Side of Things

Again, the political leaders will provide the direction on Canada's participation at COP. For the first time ever, for example, and in accordance with this direction, Canada hosted its own Pavilion at COP27 to showcase Canada's climate actions and promote dialogue among Canadians and representatives from across the world. The Pavilion sought to raise awareness of the challenges faced in Canada and the actions we are taking to help address the climate crisis. Canada had a Pavilion team who worked together with NIOs to put together the programming for the Pavilion, including its design, look, and feel. The government put out a call to all those interested in hosting a panel discussion or information session and received more submissions than there was space for. The final Pavilion programming sought out a balance among themes and representatives. While there are lessons learned we took for the next Pavilion we hosted at COP28, I knew we did a good job when a young Inuk woman, away from northern Canada for the first time, told me she felt at home when she first arrived at the Pavilion in Egypt. Canada was able to create a welcoming space for delegates to gather and share experiences.

While Canada was organizing panel discussions and programs at the Canada Pavilion, other countries and organizations were putting together their own programs and side events. The influx of invitations to the minister and officials required careful triaging, as there are always more events than people to cover them. This racking and stacking is sometimes done right up until the day of the event, as you had to be prepared for pop-up high-level events or other priority engagements where Canada needed to be present.

There is an added layer of complexity due to the fact that most COPs now tend to have a leaders segment involving heads of state and government. Not all COPs attract the same number of leaders, but a number of them have because of the significance of the issues being negotiated. Leader participation in Paris played a key role in securing the adoption of the Paris Agreement. COP26 in Glasgow also saw many leaders present to help signal continued emphasis on the need to fight climate change as the world continued to struggle with and recover from the global pandemic. In some cases, the host country lobbies hard for leader-level participation in order to raise the profile of their presidency. In other instances, the overwhelming importance of what was being agreed, such as the Paris Agreement, required the top cover of leaders. In cases where the Canadian prime minister attends, there is an increased level of planning, strategizing, and programming involved to ensure that their participation is meaningful and successful.

6.4. A Gathering That Shapes
Our Activities Throughout the Year

But COP is not about a two-week activity once per year. For example, G7 and G20 meetings can serve as important milestones that can influence COP outcomes. It is always interesting to see how much ambition there is among G7 countries to set the bar globally. The United Nations secretary-general has also advanced initiatives such as the Climate Ambition Summit in September 2023 to push on government leaders, businesses, civil society, and others to showcase actions, policies, and plans to accelerate decarbonization of the global economy and deliver climate justice. Other countries have also stepped up to help keep momentum throughout the year, including France through the launch of their "One Planet Summit" and subsequent annual gatherings.

The growth in the number, frequency, and scope of high-level events and dialogues throughout the year puts pressure on Canada's ability to be in all the places we want to be. As Canada's ambassador for climate change, I represented Canada abroad, including on behalf of ministers on a wide range of issues, advocating for our priorities, defending our interests, and maintaining our proper place at the table. And I joined a growing group of climate envoys from around the world who are similarly mandated to give increased visibility to climate change, demonstrate climate action being done at home, engage, and press others to do more.

No matter where you are in the annual COP cycle, one issue that permeates all discussions is that of climate finance and support for developing countries in addressing climate change. In 2009 at the Copenhagen COP, developed countries committed to mobilize $100 billion per year by 2020 to support developing countries in their climate efforts. The Paris decision extended this commitment out to 2025. Now that we have passed 2020, developing countries have been pressing developed countries to see the evidence of the $100 billion commitment. In 2021, Canada joined forces with Germany to provide clarity on how and when developed countries will meet the $100 billion climate finance goal. Our two countries provided updates on progress in subsequent years to help build confidence and trust with developing countries ahead of negotiations. Advocacy efforts around this important commitment continue all year at all levels.

6.5. There Is Also What We Take Away from COP, the Commitments We Make, and How We Apply Them at Home

For example, at COP28 in Dubai, Canada joined nearly 200 other countries in reaching an agreement on the global stocktake (as mandated by the Paris Agreement), which calls for tripling renewable energy, doubling energy efficiency, and transitioning away from fossil fuels in energy systems, among other things. The responsibility then falls on our planners, policy-makers, legislators, and regulators to help put these commitments into reality back home, which can take months and years to do. All eyes will be on the development of our next nationally determined contributions (NDCs), for example, and how they reflect the outcomes of the Global Stocktake.

All these activities—be they negotiations, Canada-led initiatives, the assessment of declarations and statements, ministerial briefings, Cabinet decision-making, COP programming elements—cannot be advanced without a dedicated team of public servants. Canada has been able to raise the visibility of important climate issues globally thanks to teams who create the strategies and implement them, build collaborative networks, work with our embassies and missions abroad to advance our advocacy, organize events and workshops, and generate supporting communications products. As climate change touches upon many aspects of our society, these initiatives have also required a coordinated, whole-of-government effort to advance them. I wrote about the pressures on negotiators at COP, but there is also the tireless work of the administrative team in supporting all this work, booking travel and hotels, ensuring meetings are scheduled and supported. Our IT help are the unsung heroes in this space as well—working around the clock to make sure we can do our jobs successfully from all over the world. They do the set-up and take down of our delegation room, and spend endless hours in the delegation office to ensure everyone has what they need. It is truly a Team Canada approach.

There is a considerable amount of activity required to represent our country and our climate priorities at COP. It is not just a two-week commitment. Rather, it helps to centre our activities and engagements around the globe and within Canada throughout the year. Although progress in the multilateral setting can seem slow, we have over time been able to make progress in raising the visibility of climate change and pressing each other for enhanced and more ambitious climate action. With economic opportunities in mind, we are now at the point where solutions are being brought forward and put on the table, such as coal phase-out, methane reduction, and the recognition that we need to transition away from fossil fuels in line with our 2050 net-zero goals. While the window of opportunity of keeping global warming within 1.5°C may appear to be closing, countries are all more laser focused on trying to keep this goal within our grasp given the devastating consequences that each 0.1°C of warming represents. Over the years since the first COP, the impact of climate change has become more real. Within the UN setting, small countries, including Small Island Developing States, have a strong voice and consistently remind us of the threat of climate change to their very existence. There has been recognition of the need to better support developing countries that experience loss and damage due to the impacts of climate change through the creation of a dedicated fund. And we know that we need to move to talking more about the trillions that are needed to help us address the impacts of climate change. Indeed, at COP29, parties agreed to a new collective goal to support developing countries that will scale up finance from all actors, public and private, to US$1.3 trillion per year by 2035, in addition to a tripling of the previous $100 billion goal from developed countries to $300 billion.

It is easy to get overwhelmed and discouraged in the climate space. I will never forget being stopped on the way to one of my meetings by a tearful woman who described herself as a grandmother who was very concerned about the world her grandchildren would inherit. I have met with youth representatives who are angry that not enough is being done and struggling with ecoanxiety. COPs have been that convening place to let these expressions come out, but they are also a space to showcase what is being done. Throughout the eight COPs I have attended, I have had the privilege of meeting people from all parts of society who are working hard to make a difference, and this inspires me and gives me hope. Solutions are out there. Innovations are being advanced. With the cost of climate change to society, the economic imperative and opportunities of climate action have become more and more apparent. Partnerships are working to join ideas with the much-needed policy frameworks and financing. What we have done as a global community has made a difference. And I am proud of the work of Canada's public service in helping to make that difference.

CHAPTER 7

What Now? A Millennial's Take on a Decade of COPs

Dominique Souris

> Another world is not only possible, she is on her
> way. On a quiet day, I can hear her breathing.
> *Arundhati Roy*

The United Nations Framework Convention on Climate Change (UNFCCC)—the UN convention that organizes the Conference of the Parties (COPs)—and I have a lot in common. We were both born in 1992, back when carbon levels in the atmosphere sat at 356.42 ppm.[1] We believe in the power of collaboration and consensus. And now, in our thirties, we are both forced to grapple with the existential questions—and societal expectations—that come with this age. Am I doing the right things? Am I doing things right? Thirty-plus years in, is this really what progress is supposed to look like?

I have spent over one-third of my life engaging in the COPs. My twenties revolved around them; the experiences and lessons I gained there did not just shape my career—they defined it. I started my COP journey in 2013 as an undergraduate student, during the Harper years, when the idea of a globally binding agreement was out of reach, Canada's

---

1. NASA's Goddard Institute for Space Studies.

reputation was as oily as the tar sands, and any mention of Canadian politics to international counterparts was met with a Rob Ford reference. Since then, I have spent each year engaging in one way or another in the process. Nerd? You bet. At the time of writing, I have attended nine COPs (COP19, COPs 21–28) in various roles inside, outside, and in between the formal climate negotiations. I have been a youth activist and member of YOUNGO, the official youth constituency of the UNFCCC, demanding urgency from world leaders and negotiators (COP19–21); a young researcher of global governance and finance innovation interviewing top experts on approaches to remove bottlenecks in climate negotiations and unlock climate finance (COP19–22); an advisor to the Seychelles ambassador and national delegation, creating with them one of the world's first youth negotiator programmes (COP21–25); and most recently, as an impact strategist working to advance the role of technology and philanthropy in scaling climate solutions rooted in justice (COP26–28).

Why do I keep going back? For all its shortcomings, the UNFCCC remains the only multilateral process solely dedicated to addressing the climate crisis, bringing together all sectors of society—governments, frontline communities, business, innovators, investors, scientists, creatives, and civil society. Its mission is more important than ever. Climate impacts are worsening while political will is declining. Emissions are at an all-time high, but change is happening faster than we think. We have the solutions, and the transformations we need are already underway. The COPs must play a key role in accelerating progress rather than inertia.

Perhaps like most millennials, the UNFCCC is wondering what its job really is. Thirty COPs later and the story so far is one of "agreeing to agree," rather than delivering the action, accountability, and measurable progress the world so urgently demands. A few triumphs of diplomacy met with hollow follow-through. Trust—or lack thereof—is the central tenet: between the wealthy countries responsible for the climate crisis and those most impacted, yet least responsible for causing it. COPs have become an annual stage for the former's unkept promises and platitudes, while the latter continue to fight for their survival, often pushed to concede for the sake of moving things along. As a young*ish* social entrepreneur who has spent the first decade of my career in this world, my take on the COP process is equal parts outrage and optimism. It is the lack of political and ecological integrity—or what social scientists might call "bullshit"—set against the backdrop of the positive: the *building* of new movements and unlikely collaborations, and the deep sense of *belonging* these events foster among those committed to pushing for genuine change.[2] Whatever the past has been, the future demands more. As we look ahead toward 2030 and 2050, the only story worth telling is one of bold, relentless action—one where COPs reflect the urgency and ambition of those who never stopped moving us forward.

7.1. The Beginning

I was a senior in high school during the "Great COP Out," when the world failed to land a global agreement in Copenhagen in 2009 (COP15). I remember our World History teacher telling us how world leaders had failed to seize the opportunity to make the world a better place for our generation. He was distraught as he explained that despite over 100 heads of state in attendance, the COP produced a meek political declaration—instead of the "politically binding" agreement they set out to land—which even failed to be agreed by all parties. At the time, his sense of despair and disappointment was completely lost on me; it felt so dramatic. I have now come to know this feeling after almost every single COP I have attended. But it hit hardest after the first one in 2013.

---

2. Hayley Stevenson, "Reforming Global Climate Governance in an Age of Bullshit," *Globalizations* 18, no.1 (2020): 86–102, https://doi.org/1 0.1080/14747731.2020.1774315.

The setting was a melancholic Warsaw, and the atmosphere, like the weather, was bleak. Expectations for COP19 were dismal even before it began. And if there was any doubt about where priorities lay, the Polish presidency hosted a Coal and Climate Summit on the sidelines of the formal conference. Canada, and other high-emitting countries, were retreating on their climate promises while the future of the UNFCCC was being openly questioned. I was participating as part of the University of Waterloo student delegation, one of a handful of young Canadians from across the country, representing universities, youth networks, and civil society organizations. Canada had recently left the Kyoto Protocol, so we knew the stakes were high, and we came with the intention to raise our voices and demand our government take stronger action. With support from professors and experts from the Faculty of Environment and the Waterloo Climate Institute (then known as "IC3"), we set out to engage our community and consult youth voices on campus. We arrived in Warsaw armed with a policy brief detailing key demands to present to the Canadian delegation.

Our feeling of boundless enthusiasm quickly met with the harsh realities of the COP experience. The first day in Warsaw felt alive with possibility; we were eager to engage, ready to change the narrative. But as the days dragged on, it became painfully clear that these spaces were not for us, by design. For many young people, attending COPs presents a daunting array of challenges. Securing accreditation requires knowledge, connections, and persistence to navigate a complex and opaque system, just to obtain the badge that grants access to the conference grounds. Finding enough funding for travel, accommodation (often shared and far from the venue), and basic food is another hurdle. This is even more difficult for youth from marginalized communities, especially those in the Global South, resulting in the overrepresentation of youth from the Global North in COP spaces. Once you finally arrive, you think the hard part is over, but it is just the beginning. For us in Warsaw, the few meetings we could get with the Canadian negotiating team proved sobering. Despite our preparation and enthusiasm, we had no real power—and it showed. That is often the reality of first-time COP goers; navigating a complex, intimidating environment where your presence is questioned and your voice is drowned out. Ambitious intentions often collide with the slow, closed machinery of the process, resulting in a COP experience that can feel like an uphill battle.

Thankfully, YOUNGO exists. The official youth constituency of the UNFCCC serves as a network, community, and training ground for most children and young people (defined as under 35 years old) attending the COPs. Members of YOUNGO, student groups, and organizations from around the globe self-organize to deliver capacity-building training, daily briefings on key negotiation issues, and lead working groups to articulate youth policy positions and plan actions that call attention to critical issues. For those of us privileged to come from countries with strong youth and civil society movements, the COP experience is met with additional support and community to advance national-focused priorities. COP19 was my first introduction to the Climate Action Network-Canada/Réseau action climat (CAN-Rac), which has been instrumental not only in supporting Canadians at COPs but also in strengthening climate action efforts back home.

When negotiations are at a standstill, it is young people, Indigenous leaders, civil society, and frontline communities that are the clarion call—the moral reminders who, through organizing actions, press conferences, and campaigns, demand accountability and ambition. COP19 was my first exposure to the power of organizing in UN spaces. It was also the first time I met youth from all around the world, full of passion and incredible ideas but critically underfunded. Many shared similar stories—grappling with their governments' indifference, struggling to make their voices heard, and yet pushing forward with tenacity, on a shoestring budget, if one at all. Throughout history, social movements led by young people have achieved transformative change. Imagine if they were properly supported and enabled? This was my first real glimpse into the global youth climate movement, and the moment I realized that passion was not the problem. Power and resources were. It was my "aha" moment, the one that propelled me to go further: joining YOUNGO's efforts to advocate for loss and damage financing and enshrine inter-generational equity into the Paris Agreement, and ultimately leading to the founding of the Seychelles Support Team and Youth Climate Lab.

My COP beginning made one thing clear: the narrative, the work, was bigger than any one of us. We are part of a collective struggle that demands we bridge generations and backgrounds. COP19 did not just introduce me to the process—it introduced me to the people reimagining what it could be.

7.2. Navigating the "Age of Bullshit"

Loud cheers erupted when the Paris Agreement was adopted by the UNFCCC's 196 parties on December 12, 2015—it was an undeniable diplomatic triumph and an electric moment to witness. But as I think back to COP21, it is hard to ignore the bittersweet reality of that moment. "For what it could have been, it's a miracle. For what it should have been, it's a disaster," as the British journalist George Monbiot put it.[3]

Thirty COPs later, where do we stand? It is a timeline filled with speeches, commitments, and new terminologies, yet emissions keep rising. As social science Prof. Hayley Stevenson put it, we are living in an "age of bullshit."[4] Year after year, new pledges and buzzwords dominate but the

---

3.   George Monbiot, "Grand Promises of Paris Climate Deal Undermined by Squalid Retrenchments," *The Guardian*, December 12, 2015, https://www.theguardian.com/environment/georgemonbiot/2015/dec/12/paris-climate-deal-governments-fo ssil-fuels.

4.   Hayley Stevenson, "Reforming Global Climate Governance."

reality remains the same: The climate crisis is getting worse, and the most vulnerable countries are still waiting for the finance they were promised. In these spaces, rhetoric takes precedence over ecological integrity, and the decisions made entrench the very systems leaders claim to want to transform.

I was very privileged to work with two giants of the climate world: Seychellois Ambassador Ronny Jumeau and Bangladeshi Professor Saleemul Huq, both visionary COP veterans, legendary mentors, and unapologetic champions for youth. From them, I learned the painful truth: progress in the UNFCCC has always come slow, and only after relentless pressure from those with the most to lose. It took years to even recognize that emissions had to come down. Years more to accept that adapting to climate impacts mattered too. And only after both were met with half-measures did the world begin to acknowledge what vulnerable countries and their allies had been shouting since the very beginning: that climate change was causing unavoidable losses and damages, and that this lived reality could no longer be ignored. It was in 1991 that the chair of the Alliance of Small Island States (AOSIS) first proposed the idea of compensating those most affected by the irreversible harms to their homes, cultures, and futures. It would take another 32 years for COP27 to finally establish a Loss and Damage Fund.

Climate disasters are escalating faster than the world's response, and entrenched systems and actors block every turn. This is not just a "woke" critique—the Intergovernmental Panel on Climate Change (IPCC), the world's leading authority on climate science, unequivocally links fossil fuels to the crisis.[5] Yet, at COP28 in the United Arab Emirates, fossil fuel lobbyists (2,456 of them) outnumbered the delegates from the 10 most climate-vulnerable countries combined.[6] The tide of

---

5. IPCC, *Sixth Assessment Report (AR6)* (IPCC, 2023).

6. Global Witness, "Record Number of Fossil Fuel Lobbyists Granted Access to COP28," press release, December 5, 2023, https://www.globalwitness.org/en/press-releases/record-number-fossil-fuel-lobbyists-granted-access-cop28- climate-talks/.

change was stronger; the COP28 agreement was the first COP text to mention a global shift away from fossil fuels. We celebrated it as though the mere mention of the problem was a radical act. In the world of COP, just naming the crisis, after long negotiations, is a victory. The climate crisis, however, will not pause for acknowledgements. It does not negotiate. It does not compromise. The planet continues to change regardless of what is said—or not said—in these rooms and negotiated texts.

And there is another maddening irony: While young people in these spaces are encouraged as "the solution," it is a recognition that rarely comes with real power. Almost every speech I have given at these forums—to ministers, ambassadors, negotiators, and business leaders within the UNFCCC and other UN bodies—has been with nods and the same lines: "Youth are the key!" "I have a kid, so I am serious about this too." And every time, we would respond: "Okay, then do your job." Because if you are sitting in a decision-making chair, raising awareness is the bare minimum. It is the meaningful action that remains elusive.

The status quo in climate governance allows leaders to say just enough to buy time until the next promise, the next COP. But every empty declaration corrodes trust in these spaces and undermines those working tirelessly for real change. What kind of global cooperation might be possible if these gatherings became less about self-congratulation and more about real accountability? If the mere mention of fossil fuels gave way to binding commitments to phase them out?

7.3. Building and Belonging in the UNFCCC

I see glimpses of this other world in every corner of the COPs. Indigenous and local communities lead with declarations grounded in generations of wisdom and resilience. Intergenerational exchanges, where young activists collaborate with Elders and policy influencers, connect fresh urgency with lived and deep-rooted experience. Innovators push forward bold ideas, from breakthrough technologies to grounded, community-led solutions. Artists and storytellers transform sterile conference halls into spaces of memory and meaning. In these moments, the alternative world Roy speaks of—in the epigraph to this chapter—is not just a distant dream, it is here, pulsing through each act of solidarity, each refusal to accept business as usual.

In my experience, COPs are more than just arenas for negotiation. They are sites for co-creation, where new initiatives emerge out of necessity and urgency. One of them was the Seychelles Support Team (SST), which we created as a training ground for the next generation of climate negotiators. In collaboration with the SIDS Youth AIMS Hub-Seychelles and Ambassador Ronny Jumeau, we built the SST to equip young Seychellois and youth across the Global South and North to meaningfully engage in the negotiations. It became a model of solidarity in action, built on trust, empathy, and the international cooperation too often missing in official rooms. Building on this, and recognizing the systemic gaps in support for youth, I co-founded Youth Climate Lab to enable young people to drive innovation and accelerate action in their communities and within the UNFCCC.[7] What emerges at COP does not just stand alone. It builds on the wisdom of those who came before. Youth campaigns evolve, new movements take shape, and the next generation of activists, communicators and innovators are rewriting the narrative.

COPs can help midwife the new systems and solutions that will define the future, rooted in past efforts but infused with fresh energy. For all the jargon and endless negotiations, these gatherings offer something rare: a space to seed new alliances and grow the global web of collaborators, mentors, and friends bound by a common mission. For many, it is the one time each year to reconnect in person, to feel part of

7. "Homepage," Youth Climate Lab, accessed on December 20, 2024, www.youthclimatelab.org.

something bigger. Every COP feels like a reunion—an odd, sprawling family of people from all walks of life, each bringing their expertise, urgency, and dreams to the table. Whether it is a laugh in the hallway or a long-awaited coffee with someone you have only ever known online, those small moments of connection can make all the difference. They are just a snapshot of a much larger movement, but a vital reminder that, even amid all the setbacks, we are not alone.

COPs can feel like a place to belong. But belonging is not a given for everyone. The UNFCCC is still far from safe for everyone and not accessible to all. For many racialized and marginalized people, climate spaces remain an uneven playing field, and COPs are no exception. Barriers to participation, exclusionary policies, and the erasure of intersectional identities continue to shape who is heard and who is left out. And yet, even within these flawed spaces, moments of sanctuary exist. For some queer people, especially those who cannot be "out" in their home countries, COPs can offer rare visibility, safety, and community. In recent years, I and fellow collaborators began Queering Climate, a community and collaborative studio created to celebrate and connect 2SLGBTQ+ people working on climate. Our goal has been to leverage our lived experiences and unique insights to "queer"—to challenge, question, and transform—the various streams of climate action we are involved in, whether that is policy, technology, philanthropy or beyond. At COPs 27 to 29, hosted in petro-states where homosexuality is criminalized, the Canadian Pavilion became a beacon of hope—a space to gather, to be seen, and to spotlight the intersection of LGBTQ+ rights and climate justice. These gatherings have shown that inclusion fosters innovation and creativity, enriching the climate dialogue.

Belonging is not only about finding your people—it is also about being on the right side of history. At COP21 in Paris, you could feel this in the air. No country wanted to go home empty-handed. There was a tangible hunger to be part of something historic, to belong to the side that future generations will thank. Paris was different; everyone could sense it. This was not just another round of climate talks—it was a pivotal moment. When the dust settled, and the Paris Agreement was reached, it was not simply a collection of signatures on a page; it was a profound moment of collective will, a testament to belonging to the vision of a livable future.

In the end, this sense of belonging is not just about community; it is about a deeper, shared purpose. In the face of a climate crisis that often feels insurmountable, COPs are where we come together, year after year, with the hope, and the determination, to keep pushing forward. In an era of rising geopolitical tensions and deepening uncertainty, these spaces matter more than ever. The alternative to multilateralism is not some neutral absence; it is fragmentation, division, and, in its most extreme form, war. Amid growing fractures on the global stage, strengthening the COP process is not optional. It is essential. We must ensure it is truly fit for purpose, capable of delivering the just, effective, and ambitious climate action the world demands.

7.4. Building a Future Beyond

COPs are collective pledges to the future. And that future demands implementation, integrity and ambition. What got us here will not get us where we need to go. The stakes are higher, the window is narrower, and the pace is still too slow. We need to inject new energy, urgency, and sense of purpose into these global gatherings, transforming them from the slow, line-by-line negotiations into catalysts for action.

We are on a trajectory toward 3.1°C, far from the 2°C threshold and a world away from the aspirational 1.5 °C target set in Paris.[8] That gap between aspiration and action grows wider by the day, and every delay inches us closer to irreversible tipping points. COP30 is not just another milestone; it

---

8. UNEP, *Emissions Gap Report 2024: No More Hot Air… Please!* (UNEP, 2024), https://www.unep.org/resources/emissions-gap-report-2024.

is the test we cannot fail. It took over two decades to secure the Paris Agreement, and five more years to finalize its rules. The COPs set the floor—the minimum consensus countries can agree on. But a floor is just the beginning. We need a full structure: solid, just, and ready for the storm ahead. And with climate conditions eroding, the call for structural reform is impossible to ignore. Looking toward COP35 or COP50, we have to ask: Can these forums keep pace with what is needed, or will they remain too slow, too safe, too stuck?

Until now, COPs have moved like a slow train, inching forward with small wins and cautious commitments, even as those on the frontlines and the science demand greater urgency. And while a consensus-driven approach ensures that every country has a seat and voice at the table, the incremental pace simply does not match the crisis. Without bold shifts, COPs risk either breaking down entirely or becoming mere ceremonial gatherings. So, what has to change?

The Club of Rome and other international leaders have laid out a blueprint to shift from talk to action, which includes moving from once-a-year negotiation marathons to smaller, more frequent gatherings focused on getting things done. That means redesigning how presidencies are chosen, tracking finance with real transparency, and rebalancing who gets to shape the outcomes.[9] COPs were built to be representative, but in practice, power is uneven. The voices of those most affected—Indigenous Peoples, local communities, frontline leaders—are too often drowned out by hierarchy, access barriers, and budget lines.

We know things need to be different, and we are beginning to see early signs of reinvention. One of the most exciting is the launch of a global citizens' assembly process for COP30:

---

9. Club of Rome, "Open Letter on COP Reform to All States That Are Parties to the Convention Mr. Simon Stiell, Executive Secretary of the UNFCCC Secretariat and UN Secretary-General António Guterres," https://www.clubofrome.org/cop-reform-2024/.

a bold experiment to bring the voices of everyday people—especially those on the frontlines—into climate decision-making. The idea is simple but powerful: gather representative groups from around the world to deliberate on climate priorities, then channel those insights into COP30 and beyond. It builds on earlier efforts like the *Talanoa Dialogue*, introduced by the Fijian presidency at COP23 and held throughout 2018, which used a Pacific Island tradition of open, inclusive conversation to bring governments and non-state actors into more empathetic, transparent discussions to inform countries' Nationally Determined Contributions (NDCs).

Now imagine going further. What if citizen assemblies were not one-off innovations but a core part of how global decisions get made? What if every COP had its own assembly—informing national positions, surfacing community wisdom, and holding negotiators accountable? What if the process recognised nature as a stakeholder and gave voice to those yet unborn, drawing on legal and cultural frameworks already in practice around the world?

But broadening who is heard at COPs is not enough on its own. We also need to rethink *how* these forums function. Rather than convening fragmented conversations in siloed rooms, COPs could become active hubs for collaboration—where policymakers, scientists, Indigenous leaders, youth, entrepreneurs, and artists come together not just to speak, but to build. Imagine a COP that works more like a climate innovation lab: where the policy wonks, grassroots organisers, technologists, and funders do not just pass each other in the halls, but co-design solutions that blend finance, governance, culture, and community action. Right now, the people who hold pieces of the solution are too often stuck in separate tracks—climate justice advocates here, financiers there, and technologists off to the side. But the challenges we face are inherently interconnected. So must be the solutions. COPs could be the space where those intersections are forged, where innovation is not a side event but the main agenda.

This is especially true for climate finance: if local voices, policy shapers, and investors sat around the same table, we might begin to see real accountability, transparency, and equitable access to capital, especially for those on the frontlines.

And what about Canada's role? We have had our moments, both promising and disappointing. When then-Prime Minister Justin Trudeau and Environment Minister Catherine McKenna declared "Canada is back" at COP21 in Paris, it sparked fresh optimism that Canada would be a constructive player on the world stage. Since then, we have seen our delegates, including Minister Steven Guilbeault, step up in negotiations and push for ambitious outcomes. But we have also seen promising rhetoric followed by backsliding and watered-down commitments. If Canada wants to lead—and I believe it can—it must go beyond moments. It must sustain momentum, bridge divides, and push for the structural changes that elevate those too often unheard. As a middle power, Canada has a unique role to play in brokering trust and championing transparency and accountability.

So, where does that leave us? The COPs have laid foundations, but a foundation alone is not enough. To get to a livable future, we need to build the scaffolding too: adaptive, inclusive, and strong enough to support what comes next. These gatherings must evolve from floor-setting exercises into spaces of transformation: dynamic, participatory, and bold. We have an opportunity, perhaps our last, to reimagine what these forums can be. We need a process no longer stalled by the inertia of diplomacy, but lifted by the urgency and imagination of those pushing for change.

Because in the end, the future of COPs hinges on three things: accountability, innovation, and inclusion. We must build on what exists, push further, and resist the gravity of business as usual. COPs cannot remain gatherings of minimal ambition; they must become launchpads for courageous climate action, for justice, and for the kind of world we still have time to build.

Advancing Decarbonization and Decolonization: Lessons from Indigenous Peoples Participation in the United Nations Framework Convention on Climate Change

Graeme Reed

8.1. Opening Words

For over three decades, the United Nations Framework on Climate Change (UNFCCC) has occupied the apex of international bodies working on climate action. As the largest of the three Rio Conventions (1992), a growing number of representatives from parties, Indigenous Peoples, civil society, and businesses migrate, nearly annually, to locations around the world to advocate for more robust action on the climate crisis. Recently, for example, the United Arab Emirates (UAE) hosted the twenty-eighth session of the Conference of the Parties where over 85,000 participants attended, 34,000 more than the preceding year in Egypt, and nearly 30,000 more than the preceding year in Azerbaijan. Over the course of these three decades, discussions within the UNFCCC system have transitioned through different negotiated agreements, including the Kyoto Protocol, the Copenhagen Accord, and the Paris Agreement. The Paris Agreement, ratified in 2016, is a universal agreement to hold "the increase in the global average temperature to well below 2°C above pre-industrial levels and pursuing efforts to limit the temperature increase to 1.5°C."[1]

Despite the growth in participation of these annual conferences, collective emission reduction trajectories trend in the opposite direction. The 2024 *Nationally Determined Contributions (NDCs) Synthesis Report*, produced by the UNFCCC Secretariat, confirmed that the full implementation of current NDCs, including their conditional elements, will increase emissions by 5.9 percent by 2030, rather than decrease them.[2] The implications of these findings are outlined in the 2024 UN Environment Program's (UNEP) *Emissions Gap Report*, indicating that we are on track to global warming of up to 2.6°C by the end of the century.[3] A decrease of 43 percent, according to the Intergovernmental Panel on Climate Change (IPCC), is required to avoid locking in substantive adverse climate impacts and losses and damages. Even over the past two years, climate-induced extreme weather events have been on the rise in Canada, with, for example, floods in Québec City, Toronto, and Vancouver, as well as out-of-control fires burning over 15 million hectares across the country. In the face of these rising climate impacts, one would think that the kind of "deep, rapid and sustained" emissions reductions required by the IPCC would be forthcoming. Yet the opposite is true—emissions continue to rise year after year. Based on my active participation in UNFCCC

1. *Paris Agreement*, 12 December 2015, 3156 UNTS 79 (entered into force 4 November 2016), art 2(a) [*Paris Agreement*].

2. The full report can be found here: UNFCCC, *Nationally Determined Contributions Under the Paris Agreement: Synthesis Report by the Secretariat* (UNFCCC, 2024), https://unfccc.int/sites/default/files/resource/cma2024_10_adv.pdf.

3. For more information, see: UNEP, *Emissions Gap Report 2024: No More Hot Air … Please!* (UNEP, 2024), https://www.unep.org/resources/emissions-gap-report-2024.

sessions since 2017, I will share in this chapter how this incongruence is the outcome of a process that was designed to knit an increasingly complex quilt of geopolitical horse-trading, techno-optimism, and false solutions, rather than address the root causes driving the climate crisis. Root causes that Indigenous Peoples have been identifying internationally since 1992 during the Declaration of Kari-Oca.[4]

To do this, I use the emergence of the Facilitative Working Group (FWG) of the Local Communities and Indigenous Peoples Platform ("the Platform") to discuss the evolution of Indigenous Peoples' participation within the UNFCCC. As the only UNFCCC constituted body that has equal representation from Indigenous Peoples and state representatives, reflecting on the history of the Platform can provide valuable lessons on how Indigenous Peoples participation, worldviews, and advocacy can unravel the quilt to focus on the root causes of the climate crisis. I will discuss this evolution using my first-hand experience in the creation and implementation of the constituted body as a party delegate (Canada), a co-chair of the Indigenous Peoples Constituency, and as the Indigenous North American member of the FWG. In what follows, I introduce my positionality and the experience of Indigenous Peoples within the UNFCCC, and then shift to a substantive discussion of the negotiation process to create the FWG, including the Government of Canada's central role. I then explore what the presence of the FWG, and Indigenous Peoples more broadly, mean for the current and future state of the UNFCCC, concluding with some reflections about its potential role in advancing both decarbonization and decolonization.

---

4. *Declaration of Kari-Oca*, May 30, 1992, cwis.org/wp-content/uploads/documents/kari-oca.txt.

8.1.1. How I Enter This Space

My academic work, and experience working at the Assembly of First Nations (AFN), have shown me the importance of writing in "a good way." For me, this means approaching it using an Indigenous research paradigm (IRP), working on "being in" Indigenous sovereignty to deconstruct the dominant assumptions within climate discussions to advance Indigenous solutions grounded in Indigenous Knowledge, legal orders, and governance. As such, it is appropriate for me to first introduce myself. *Aanii*. Graeme Reed *nindizhinikaaz*. Ottawa *nindoonjibaa*. I was born and raised in Ottawa with mixed ancestry from England, Scotland, Germany, and mnidoo-gaamii Anishinaabe (Georgian Bay). My great-grandfather was born in Wiikwemkoong Unceded Territory on Odaawaa-minis (Manitoulin Island). For the last eight and a half years, I have worked at the AFN leading our domestic and international climate work, including participating in the UNFCCC and the IPCC.

I still remember the feeling of arriving at the 46th Subsidiary Session[5] (SB46) in Bonn, Germany for the first time in 2017. A mixture of anxiety, excitement, and uncertainty collided with an unexpectedly warm spring day (I will let you infer what my first photo looked like), as I walked into the World Conference Center: A space that has become both familiar and deeply problematic. Since then, I have participated as a representative of the AFN on the Canadian delegation at every intersessional, always in Bonn, and every COP since the 23rd session (COP23), hosted by the Fijian presidency, also in Bonn. The one exception to this was in

---

5. The Parties to the UN Framework Convention on Climate Change meet twice a year, once in June, and another in November/December. The June meeting is a meeting of the two subsidiary bodies, the Subsidiary Body on Science and Technological Advice and the Subsidiary Body on Implementation, where decisions get forwarded to the November session, the yearly Conference of the Parties.

June 2022, where much to the relief of my darling fiancé, I chose our wedding over the prospect of spending an additional two weeks away. As a pink-badge holder (a representative of a party), I have been afforded a unique vantage point through which to participate in COP meetings, having the potential to both participate in meetings with the Canadian negotiation team, to observe negotiations and contribute to negotiation positions, and to actively participate in the Indigenous Peoples Caucus.[6]

In 2019 at the Bonn intersessional, it became apparent that the North American Indigenous Region was next in line to nominate a co-chair for the Indigenous Peoples Caucus, also known as the International Indigenous Peoples Forum on Climate Change (IIPFCC).[7] With the helpful counsel (*read* volun-telling) of several Knowledge Keepers and longstanding participants within the Indigenous Peoples Caucus, and the approval of my supervisor back home, I was nominated for the position on an interim basis. Perhaps I was naïve, but I did not realize how much responsibility this position carried, thrusting me into four years of representing the Caucus in high-level discussions with multiple COP presidents (and their teams), the UN secretary-general, and the UNFCCC executive secretary; organizing five consecutive Indigenous Peoples Pavilions; and convening Indigenous

---

6. It is worth noting that the Government of Canada is quite flexible in their approach to Indigenous Peoples participating on the delegation. While there are protocols to follow, it does not restrict our participation and voice, except if attempting to speak on behalf of Canada, or if disclosing confidential information. For me, this is just common sense, so it has not been a problem. The Indigenous Peoples Forum on Climate Change (IIPFCC), also known as the Indigenous Peoples Caucus, is one of nine officially recognized constituencies of accredited civil society organizations within the UNFCCC.

7. The Indigenous Peoples Caucus is made up of seven socio-cultural regions: Africa, the Arctic, Asia, Central and South America and the Caribbean, Eastern Europe and the Russian Federation, Central Asia and Transcaucasia, North America, and the Pacific.

Peoples preparatory and daily Caucus meetings at both the intersessionals and at the COPs. I stepped down formally in 2023, though the responsibility continued as I have shifted to an advisory role for the current co-chairs.

Since 2021, I have been actively involved in the FWG, first as an alternate member for the North American region when Grand Chief Willie Littlechild stepped down, and second as the main member for the North America Indigenous sociocultural region. As you will learn through the course of this chapter, the creation of the FWG was an important step forward for the participation of Indigenous Peoples within the UNFCCC, as the only body within the entire UN system that allows Indigenous Peoples to self-select their representatives. After an intensive criteria and selection process, Andrea Carmen (Yaqui Nation), executive director of the International Indian Treaty Council (IITC), was nominated as the first representative for our region. Following in her footsteps, I was nominated by the North America Working Group on Climate Change alongside Chris Honahnie (Hopi/ Diné) for a three-year team. Early in the tenure, unfortunately, Chris tragically passed away and was replaced by Janene Yazzie (Diné).[8] As I am writing this, I am ending my term as vice-chair of the FWG and formally concluding the process for the nomination of the next North American FWG member.

8.1.2. *Indigenous Peoples within the UNFCCC*

As a result of these positionalities, I have come to intimately understand the emergence and advocacy of Indigenous Peoples within the UNFCCC. In fact, Indigenous Peoples have been participating in international discussions on climate change since before the original Rio Declaration in

---

8. International Indian Treaty Council, " A Young Warrior for Mother Earth Has Returned Home," December 31, 2022, https://www.iitc. org/a-young-warrior-for-mother-earth-has-returned-home/.

1992 (the year of my birth), passing their own declaration, the Kari-Oca Declaration. Even in the Rio Declaration on Environment and Development, parties acknowledged the unique contributions of Indigenous Peoples: "Indigenous people and their communities [...] have a vital role in environmental management and development because of their knowledge and traditional practices."[9] Despite this long-standing participation, the Indigenous Peoples Caucus was formally recognized in 2000 as one of nine formal constituencies within the UNFCCC with specific rights to participate as observers of negotiations.

To provide structure for the constituency, Indigenous Peoples organized themselves into the IIPFCC in 2008. The IIPFCC helps coordinate attending Indigenous Peoples to discuss priorities, negotiate items, and hold side events in a culturally safe space (the Indigenous Peoples Pavilion). The IIPFCC aims to unify and amplify the voices of Indigenous Peoples while maintaining that individual organizations at the subnational, national, and global levels have their own agendas, priorities and proposals to advance at the UNFCCC. Since 2008, and long before the Paris Agreement in 2015, parties to the UNFCCC have been considering the role of Indigenous Peoples and their knowledge systems in global climate action. This has been represented by more than 60 decisions adopted by COPs, or in reports adopted by subsidiary bodies that explicitly reference Indigenous Peoples and traditional knowledge.[10] At COP21 in 2015, the Paris

---

9. You will notice that these original references do not capitalize "Indigenous Peoples." This was a long-fought battle by those Indigenous leaders within the negotiations on the UN Declaration on the Rights of Indigenous Peoples.

10. The IIPFCC is currently (December 2024) working on an updated version of this report, which we expect to be released soon. The previous versions can be found here: Lisa Kadel and Sébastien Duyck, eds., *Indigenous Peoples and Traditional Knowledge in the Context of the UNFCCC* (Center for International Environmental Law, 2021).

decision text (1/CP.21) maintained this momentum, referencing Indigenous Peoples or traditional knowledge six times. Central to these references was a recognition that climate change poses a considerable threat to the realization of human rights, especially the rights of Indigenous Peoples: the Paris Agreement stressed that parties "should, when taking action to address climate change, respect, promote, and consider their respective obligations on human rights," including the rights of Indigenous Peoples. This rights-based recognition has been the foundation for advocacy in the creation of the FWG, and to broader involvement of Indigenous Peoples within the UNFCCC.

8.2. Breathing Life into the FWG and the Platform

Reflecting on the growing space of Indigenous Peoples within the UNFCCC, there is no better example than the creation, and operationalization, of the Facilitative Working Group of the Local Communities and Indigenous Peoples Platform (a mouthful). While this is currently seen as an important procedural step forward for the participation of Indigenous Peoples, it did not come without its challenges. In fact, the negotiation took numerous years, strong Indigenous advocacy at formal and informal meetings, and significant political pressure. In this section, I focus on the productive role that Canada, alongside representatives of First Nations, Inuit, and Métis, played in breathing life into the FWG, building "institutional" credibility for Indigenous Peoples and shepherding a new era of participation for Indigenous Peoples at the UNFCCC. However, it may be helpful to provide a high-level overview of what exactly the Platform is, and the role of the FWG.

The origin of this body can be traced to the final decision of COP21 (1/CP.21) on the adoption of the Paris Agreement, in its paragraph 135: "*Recognizes* the need to strengthen knowledge, technologies, practices and efforts of

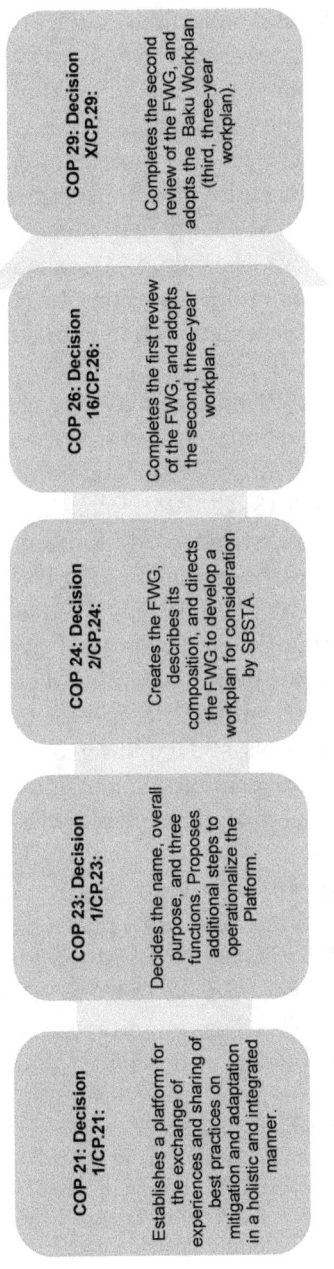

COP 21: Decision 1/CP.21:

Establishes a platform for the exchange of experiences and sharing of best practices on mitigation and adaptation in a holistic and integrated manner.

COP 23: Decision 1/CP.23:

Decides the name, overall purpose, and three functions. Proposes additional steps to operationalize the Platform.

COP 24: Decision 2/CP.24:

Creates the FWG, describes its composition, and directs the FWG to develop a workplan for consideration by SBSTA.

COP 26: Decision 16/CP.26:

Completes the first review of the FWG, and adopts the second, three-year workplan.

COP 29: Decision X/CP.29:

Completes the second review of the FWG, and adopts the Baku Workplan (third, three-year workplan).

Figure 8.1. A Short Overview of the Platform.

local communities and indigenous peoples related to address-ing and responding to climate change, and establishes a plat-form for the exchange of experiences and sharing of best practices on mitigation and adaptation in a holistic and inte-grated manner." The process to respond to this statement involved a series of multi-stakeholder open dialogues (the first of which was at SB46, my first intersessional), informal workshops in Ottawa, Helsinki, and Cochabamba, and regu-lar negotiations over multiple sessions. In 2018 (COP23), the first step towards operationalization was taken, where parties agreed to the name of the body, identified its three main func-tions (knowledge exchange, capacity for engagement, and cli-mate policies and action), created a process to establish the FWG and the modalities for the creation of a work plan.[11] The next year, in 2018 (COP24), following several informal workshops and sustained advocacy, Indigenous Peoples were successful to further operationalize the Platform. The deci-sion created the FWG as a body with equal representation between Indigenous Peoples and states (seven each), includ-ing in leadership positions, enabling Indigenous Peoples to self-select their representatives according to their respective processes and protocols, and empowered them to develop an initial two-year work plan during one of their two biannual meetings. The initial two-year work plan was developed at the SB session in June 2019 and adopted at COP25 in Madrid. Following two years of virtual activities over the COVID pan-demic, a second three-year work plan was developed and adopted at COP26 in Glasgow. Most recently, a third three-year plan has been adopted at COP29 in Baku, where it dealt with the addition of local community representatives, among other issues. For the sake of brevity, additional information

11.   For the full decision, Decision 2/CP23, refer here: UNFCCC, *Report of the Conference of the Parties on Its Twenty-Third Session, Held in Bonn from 6 to 18 November 2017. Addendum. Part Two: Action Taken by the Conference of the Parties at its Twenty-Third Session* (UNFCCC, 2018), https://unfccc.int/documents/65126.

(and nuance) about these steps can be found in chapters of the Indigenous World from 2019 to 2024, a publication produced by the International Working Group on Indigenous Affairs (IWGIA);[12] and the details about the Platform's work plan can be found on their website.[13]

8.2.2. Canada's Leadership Role in the Negotiation of the Platform

It cannot be understated the role that friendly parties, such as Canada, played in supporting Indigenous Peoples' leadership and visions within the negotiations of the Platform. This is not to diminish the leadership, organization, and advocacy of Indigenous Peoples from the seven sociocultural regions, but it is to acknowledge that in a "party-driven" process, support from parties is essential to move these priorities forward. Support from the "friends of the Platform" pushed the negotiating process to take important procedural steps forward for the participation of Indigenous Peoples. All negotiations—including the "informal informal" sessions—were open to Indigenous Peoples and non-party observers.[14] Through the course of the two-year negotiation leading to the creation of the Platform (2017–2019) and the four-year negotiation for the continuation and adoption of updated workplans

12. For more, refer here: IWGIA. n.d. "The Indigenous World 2024," https://iwgia.org/en/resources/indigenous-world.html. I should caveat this with the fact that I have been writing these summaries with other Indigenous representatives for the past five years.

13. For more, refer here: "United Nations Climate Change Local Communities and Indigenous Peoples Platform," United Nations Climate Change, accessed on July 2, 2025, https://lcipp.unfccc.int/.

14. Informal informals are a negotiation that do not involve the designated co-facilitators. These have increasingly been used by parties to move negotiation items forward in a less-formal setting, however, representatives of observers (both Indigenous Peoples and members of civil society) have been concerned with their frequency given they are often closed to observers.

(2019–present), party negotiators frequently called on a representative of Indigenous Peoples to provide substantive input on the negotiation text. This was particularly important in the early stages of the negotiation, as Indigenous representatives pushed for the inclusion of the rights of Indigenous Peoples, Indigenous Knowledge systems, and the UN Declaration on the Rights of Indigenous Peoples (UN Declaration). The advocacy was so successful that principles prepared by the IIPFCC were adopted verbatim in an operative paragraph in COP Decision 2/CP23 (paragraph 8). At the closing of the high-level segment at COP23, a Māori representative, India Logan-Riley, described the enhanced participation of Indigenous Peoples as reflective of "new fire, a fire of willingness and inclusion, cooperation between Indigenous peoples and parties in good faith [...]"[15] (as cited in Reed & Sadik, 2018).

8.2.2.1. Creating the Facilitative Working Group

In advance of the first multi-stakeholder dialogue in 2017, an open call for submissions was released to propose elements for the scope, governance structure, and functions of the Platform. Parties, Indigenous Peoples, UN-agencies, and allied organizations alike submitted submissions to communicate the elements that they felt would be appropriate to breathe life into the Platform and fulfill the aspirations of Indigenous Peoples. Submissions from Indigenous Peoples were largely consistent, prioritizing the rights of Indigenous Peoples, the ethical engagement with Indigenous Knowledge, and the full and effective participation of Indigenous Peoples, with some variation about the approach to the structure.[16]

---

15.   G. Reed and T. Sadik, *Operationalizing the Local Communities and Indigenous Peoples Platform: A Step in the Right Direction?* (Centre of International Governance and Innovation, 2017).

16.   For example, refer to the submission prepared by the International Indigenous Peoples Forum on Climate Change here: International

Parties, however, were widely divergent: Some proposed a simple web portal, while others proposed the potential to create a body akin to the Ad Hoc Open-Ended Working Group on Article 8(j) of the Convention on Biological Diversity (now a permanent subsidiary body). This divergence manifested in the first multi-stakeholder dialogue, creating a wide number of perspectives between parties, which left the participants uncertain about the future of the negotiations. Some overnight discussions between the head of Canada's delegation and senior leadership back home (and some pressure from the Indigenous representatives in attendance) led to a commitment from the Government of Canada to host the first of several informal meetings in Ottawa in September 2017. Additional meetings were hosted in Helsinki (February 2018) and Cochabamba (October 2018) with the intention of mapping out potential options to arrive at a place of consensus for the new program. Having participated in all three of these sessions, each enabled the process to take important steps forward in addressing key sticky items within the negotiations,

Indigenous People's Forum on Climate Change, *Submission of the International Indigenous Peoples' Forum on Climate Change (IIP-FCC) on the Purpose, Content and Structure for the Indigenous Peoples' traditional knowledge platform, 1/CP.21 paragraph 135 of the Paris Decision,* https://unfccc.int/files/parties_observers/submissions_from_observers/application/pdf/865.pdf; and a complementary version prepared by the Arctic, Latin America, and Pacific here: *The Indigenous regions of the Arctic, Latin America and the Caribbean and the Pacific submit the following comments and proposals combined in this joint submission to theUNF CCC, with regard to the Purpose, Content and Structure for the Local Communities and Indigenous Peoples platform established by decision 1/CP.21 paragraph 135 of the Paris Agreement,* 2017, https://unfccc.int/files/parties_observers/submissions_from_observers/application/pdf/871.pdf. AFN's submission can be found here: *Submission of the Assembly of First Nations (AFN) on the Purpose, Content and Structure for the Indigenous Peoples' traditional knowledge platform, 1/CP.21 paragraph 135 of the Paris Decision,* https://unfccc.int/files/parties_observers/submissions_from_observers/application/pdf/882.pdf.

such as the consideration of the participation of "local communities" within this new Platform.

In the final hours of the negotiations at COP23 in 2017, the term "Facilitative Working Group" emerged, seeking to build consensus between the divergent visions of different parties. Despite this new term, there was no certainty of what it meant in detail, or whether it was possible to create a new constituted body with equal membership between representatives from parties and "non-parties" (Indigenous Peoples). Research commissioned by the Secretariat explored the existence of similar bodies within the UNFCCC, finding that while there were groups and committees that engage with non-party stakeholders, some even including Indigenous Peoples, none included participation on the body itself.[17] In the lead-up to the negotiations at COP24 in 2018 (where the FWG was adopted), the potential of creating a new constituted body with "non-party" stakeholders was actively challenged by members of the UNFCCC Secretariat. Sitting in the dining area of the World Conference Center in Bonn, I have vivid memories of a bilateral meeting between Canadian negotiators, myself, and UNFCCC Secretariat staff (including their legal counsel), who were adamant that the vision of Indigenous Peoples to have a body with equal representation was impossible, while at the same time indicating that—given the UNFCCC functions as a party-driven process—the Secretariat would implement anything that was determined by consensus between the parties. Feeling slightly deflated, the Canadian negotiators and I took it upon ourselves to push this consensus forward, and later in that same year, at COP24 in Katowice, parties agreed by consensus to create the FWG:

17. For more information, refer to the report prepared by the Ecologic Institute, here: Arne Riedel and Ralph Bodle, *Local Communities and Indigenous Peoples Platform – Potential Governance Arrangements under the Paris Agreement* (Nordic Council of Ministers, 2018), https://www.ecologic.eu/sites/default/files/publication/2018/2139-local-communities-and-indigenous-people-platform_0.pdf.

the only constituted body with equal representation between Indigenous Peoples and party representatives. To this day, there is no similar constituted body within the UNFCCC system, and no body within the UN system that allows self-selection by Indigenous Peoples.

8.2.2.2. Trilateral Negotiations to Protect the UN Declaration on the Rights of Indigenous Peoples (UN Declaration)

During this time, Indigenous Peoples were advocating for the preambular rights-based language of the Paris Agreement (parties' obligations to safeguard human rights and the rights of Indigenous Peoples) to become operative text, meaning it would have more legal force under international law. While there was general support for the inclusion of the rights of Indigenous Peoples, a significant amount of time and attention were required to expand the inclusion of this rights-based language towards the inclusion of the wider set of rights and their respective terminology as already recognized under the UN Declaration. During the intersessional negotiations in 2018, one sticky issue emerged that threatened the very creation of the FWG, a desire from certain parties to include the protection of territorial integrity and political unity in response to the inclusion of the UN Declaration terminology in the decision text. This cherry-picking of text within the UN Declaration led to a large disagreement in the negotiation room, resulting in the negotiations being punted to COP24 in Katowice. There, the proposed watering down of the UN Declaration was a major concern for Indigenous Peoples, including representatives of First Nations, Inuit, and Métis on Canada's delegation. Thankfully, Canada took a hard stance on the exclusion of territorial integrity, believing it was already captured within Article 46 of the UN Declaration, isolating itself from other friendly parties who did not have Indigenous Peoples on their delegation. With the leadership of the head of delegation and the lead negotiator, this pressure manifested into an intense trilateral negotiation between the chair of the G77+China, the party pushing for the inclusion of territorial integrity, and Canada in the late evening of the final nights of negotiations.[18] Thanks to constant advocacy and discussion, the compromised position landed in Decision 2/CP23, where "territorial integrity" is only referenced with regards to functions involving local communities, whereas the UN Declaration in its entirety is referenced with regards to functions involving Indigenous Peoples. This would not have happened without the support of Canada in the negotiations.

8.2.2.3. Addressing the Terminology "Indigenous Peoples and Local Communities"

The final sticky issue, as evidenced by the Platform's name, is the inclusion of "local communities." The main advocates for the inclusion of local communities, originally, were parties who either did not recognize the Indigenous Peoples of their country (such as Indonesia) or those that have a specific history of traditional local communities, such as Afro-descendant Maroon and Quilombola communities in Latin America. At the same time, other parties challenged the inclusion of local communities because of the absence of any clear international definition.[19] Indigenous Peoples, who are an organized constituency within the UNFCCC, felt

---

18. The G77 was founded in 1964 in the context of the UN Conference on Trade and Development (UNCTAD) and now functions as one of the voting blocs throughout the UN system. As of May 2014, there are 135 members in the group, though many parties also participate in other party groupings and sub-groupings (such as the African Group, for instance).

19. For a more detailed exploration of the emergence of the term local communities, and some of the challenges that Indigenous Peoples face, refer to: Assembly of First Nations, *AFN Position Paper on the Terminology "Indigenous Peoples and Local Communities"* (AFN, 2024), https://afn.bynder.com/m/731677d1e91cd5b3/original/AFN-

challenged to respond to this pressure, not wanting to over-step the self-determination, self-representation, and self-mobilization of local communities but needing to protect their distinct status and rights, as reaffirmed in the UN Declaration. Despite this, parties exerted a significant amount of time and attention to reflect the inclusion of local communities within the work of the Platform, including in the hosting of specific workshops for their enhanced participation and the potential addition of up to three additional local community representatives on the FWG. The addition of local community representatives was a major concern for Indigenous Peoples, both due to the initial commitment to equal status between party and Indigenous Peoples' representatives, as well as due to the conflation between Indigenous Peoples and local communities, which poses a slow erosion of the distinct rights and status of Indigenous Peoples. This was communicated in the Indigenous Peoples Caucus Opening Statement of COP28 (2023): "our inherent, distinct, internationally-recognized rights are affirmed in the UN Declaration on the Rights of Indigenous Peoples. We will not allow these rights to be diminished, undermined, combined or confused in any way."

A slow trickle of representatives of local communities, mostly from Latin America, has come to participate within, and contribute to, the FWG meetings. Despite this, there remains no formalized constituency, no definition, nor a process of nomination for local communities' representatives to be added to the FWG. As a result, parties felt that it was too early during the first review of the FWG at COP26 to add new representatives and suggested that the matter be reconsidered at the next review, concluding at COP29. In the intervening time, international Indigenous advocacy on the conflation issue has grown, leading the UN mechanisms on Indigenous

Position-Paper-on-the-Terminology-Indigenous-Peoples-and-local-communities.pdf.

Peoples (the Permanent Forum on Indigenous Issues, the Expert Mechanism on the Rights of Indigenous Peoples, and the Special Rapporteur on the Rights of Indigenous Peoples) to issue a joint statement.[20] This statement was brought to the recent negotiations on the second review of the FWG, where Indigenous Peoples pushed to close the discussions around the addition of three local community representatives. Much to the surprise of those of us in the room, as well as bilateral advocacy between Canada and other countries, parties did not push for their inclusion, focusing instead on procedural and financial elements of the decision. The text has now been adopted at COP29, thus closing the chapter of additional "local communities" representatives in the FWG. The logical extension of this decision is how to consider this updated governance structure within the context of the full name of the Platform—the Local Communities and Indigenous Peoples Platform—and whether there needs to be modifications at the next FWG review in 2027.

Without this close partnership between Canadian negotiators and representatives of First Nations, Inuit, and Métis, many of these challenges would have prevented the full potential of the FWG to amplify the voices, knowledge systems, and rights of Indigenous Peoples. Recognizing this, we now turn to the closing section in this chapter to explore the implications of these discussions for the UNFCCC, for the future of COPs, and, frankly, for the future of life for people and for the planet.

---

20. The full statement from the three mechanisms can be found here: United Nations Expert Mechanism on the Rights of Indigenous Peoples, *Statement by the UNITED NATIONS Permanent Forum on Indigenous Issues Special Rapporteur on the Rights of Indigenous Peoples, and the Expert Mechanism on the Rights of Indigenous Peoples*, 2023, https://www.ohchr.org/sites/default/files/documents/issues/indigenous peoples/emrip/Statement_EMRIP_July_2023.pdf.

8.3. Lessons from the Creation of FWG for the Future of the UNFCCC

The rising presence of Indigenous Peoples within discussions on the UNFCCC, as represented by the development of the FWG and Platform, offers new prospects for the future, and effectiveness of the international body. To get there, however, the UNFCCC must internalize these lessons and make structural modifications to the conceptual and procedural approach to discussions on climate change. In this final section, I set out three calls to action: i) identify the root cause driving climate change; ii) widen participatory rights of Indigenous Peoples within UNFCCC discussions; and iii) ethically and equitably engage with Indigenous Knowledge systems. I address each in turn.

The reality of the "polycrisis" facing humans and the natural world has demonstrated that a siloed approach to climate change is no longer fit for purpose. The global strategy for addressing the climate crisis has historically been a narrowly focused one on the quantitative reduction of greenhouse gas emissions at all costs. To address this reality, the UNFCCC must evolve in its conceptualization of the root causes of the climate crisis. Indeed, the FWG's recent report to the Subsidiary Body for Scientific and Technological Advice (2024) identified this as a key challenge: "a fundamental misalignment between the prevailing global approach to addressing climate change and the perspectives of Indigenous Peoples […] on the changing climate."[21] Throughout the development of the FWG, and the advocacy of Indigenous Peoples leading up to it, this misalignment has been omnipresent, demonstrating that Indigenous Peoples, while diverse, problematize the drivers of the climate crisis differently than mainstream systems. To illustrate this, the FWG report

---

21. *Report of the Facilitative Working Group of the Local Communities and Indigenous Peoples Platform*, 60th sess, UN Doc FCCC/SBSTA/2024/1, 2024, p. 11.

quotes a knowledge holder who participated in the Annual Gathering of Knowledge Holders at COP28: "Humans need to be humbler when interacting with the natural world. We cannot stop sunrise or sunset, nor can we impact winter or the winds. We need to avoid human exceptionalism and stop neglecting nature, whose laws are stronger than those laws of humans."[22] Indigenous Peoples question the centrality of technological and market-based solutions without a critical investigation of the inequitable and structurally racist ways that these solutions interact with lived realities, furthering harm to Indigenous Peoples and contributing to a new form of climate colonialism. A more meaningful approach would begin with establishing a shared understanding of what is driving the climate crisis, and based on that shared understanding, determine appropriate actions. Elder Courchene-baa calls for re-evaluation of the framing of climate change towards one focused on how human values are at the root of the climate crisis: a world out of balance.[23] Moving forward, the UNFCCC must heed these calls and introduce new ways to dialogue on the root causes of the climate crisis.

The FWG is a positive step towards the enhanced participation of Indigenous Peoples within the UN system. As

22. To find the past co-chair summaries, refer here: COP 28: Local Communities and Indigenous Peoples Platform (LCIPP), *Draft Summary Report Third Annual Gathering of Knowledge Holders,* 2024, p. 11–12, https://lcipp.unfccc.int/sites/default/files/2024-11/Third%20Annual%20Gathering%20of%20Knowledge%20Holders%20at%20COP%2028-Co-Leads%20Summary%20Report.pdf, and COP27: Local Communities and Indigenous Peoples Platform (LCIPP), *Summary of Second Annual Gathering of Knowledge Holders,* https://lcipp.unfccc.int/sites/default/files/2023-12/Co-Leads%20Summary%20of%20LCIPP%20Annual%20Gathering%20of%20Knowledge%20Holders.pdf.

23. Laura Cameron, Dave Courchene, Sabina Ijaz, and Ian Mauro, "'A Change of Heart': Indigenous Perspectives from the Onjisay Aki Summit on Climate Change," *Climatic Change* 164, no 3 (2021): 43, https://doi.org/10.1007/s10584-021-03000-8.

the first constituted body with equal representation of states and Indigenous Peoples, the institutional credibility created by the FWG must be used to continue advancing the aspirations of Indigenous Peoples to participate as self-determining nations, in line with the minimum standards contained with the UN Declaration. Elder Francois Paulette, a Denesuline Elder, hereditary chief and spiritual leader from Denendeh, spoke at COP23, following the adoption of the first decision related to the Platform, to the importance of including Indigenous Peoples as "one hundred percent participants" in the UNFCCC. While this is particularly challenging in a party-driven process, the FWG has been leveraged to include Indigenous Peoples and their views on Indigenous sovereignty, rights, and knowledges, in other technical discussions at the UNFCCC, such as the Warsaw International Mechanism for Loss and Damage, advancing a "subtle revolution" (a term drawn from Sheryl Lightfoot's 2016 book *Global Indigenous Politics: A Subtle Revolution*). To take a step further and honour the commitment to the UN Declaration, our collective work on climate change must prioritize, empower and support Indigenous Peoples' full and effective participation. Doing so includes allowing Indigenous Peoples to define what "full and effective" participation means within the context of the UNFCCC and avoid the misrepresentation of the FWG as an "one-stop-shop for all things Indigenous" within the UNFCCC. Instead, enhanced participatory rights and the growing recognition of Indigenous Peoples' contributions to global efforts on climate action must be a core element of structural modifications within the UNFCCC moving towards COP30 and beyond.

There is growing international recognition of the unique role Indigenous Knowledge systems play in the creation of solutions, however, there is little engagement with what this would mean substantially. For example, based on an understanding grounded in Indigenous Knowledge systems that we are "one with the Land and Water"—rather than compartmentalized units apart from the natural world—all discussions would need to centre on the reciprocal relationships that embody our global ecosystem. This concept of "Mother Earth" while included in the Paris Agreement, if taken seriously within climate discussions, could be an important opener to move away from the mirage of scientific objectivity, with decades of scholarship unpacking and contesting this,[24] towards one that engages with diverse ways of knowing and being. This includes overcoming a knowledge hierarchy that prioritizes natural sciences, engineering, and techno-economics over social sciences, humanities, and critical approaches. Practically, the institutional arrangements of the Platform could be leveraged to facilitate and support the development of relevant mechanisms and ethical protocols for the ethical and equitable engagement with the values of Indigenous Peoples and their knowledge systems.

8.4. Conclusion

The UNFCCC remains an important institution in the global efforts to combat catastrophic climate change, despite ongoing challenges with its success and effectiveness. Using the negotiation and now implementation of the FWG of the Platform, I identified three calls to action that support the UNFCCC in shifting towards an institution that can engage meaningfully with both decarbonization and decolonization: engaging with the root drivers of the climate crisis, supporting enhanced participation of Indigenous Peoples, and ethically engaging with Indigenous Knowledge systems. The ultimate success of the FWG, and the Platform more broadly, will depend on how well it can amplify these voices and speed up transformative change amid the real, immediate, and existential threats that First Nations and other Indigenous Peoples are facing daily across Canada and internationally.

24. See, e.g., Sandra Harding, *Science and Social Inequality: Feminist and Postcolonial Issues* (University of Illinois Press, 2023).

CHAPTER 9

ᓯᓚ ᐊᑦᑎᐅᔪᓐᓇᐃᖅᑐᖅ—
Sila Ajjiujunnaiqtuq—
The Weather Has Changed:
Inuit Perspectives and Experiences from
30 Years of Climate Change COPs

*Lisa Qiluqqi Koperqualuk, Sara Olsvig,
Piita Irniq, Alexina Kublu, Dalee Sambo Dorough,
Sheila Watt-Cloutier, Miyuki Qiajunnguaq Daorana,
Susie-Ann Kudluk, and Anne Simpson*

Founded in 1977 by the late Eben Hopson of Barrow, Alaska, the Inuit Circumpolar Council (ICC) has flourished and grown into a major international Indigenous Peoples organization representing approximately 180,000 Inuit of Alaska, Canada, Greenland, and Chukotka (Russia). While this chapter is organized and put together by ICC, it reflects personal views and experiences of delegates and leaders who have taken part in the United Nations Framework Convention on Climate Change Conference of Parties (COP) and is not an official ICC publication.

Inuit of the Arctic, we are observers of the weather. We are observers of the weather changes in the Arctic, we are observers of the sky, the clouds, the ice and the land, as well as the animals that we hunt to survive. It is our culture.

–Elder Piita Irniq

9.1. Inuit Voices and the Fight for Climate Justice

Inuit in Inuktitut means "the people." This will be the focus of this chapter, which incorporates contributions from past, present, and future Inuit leaders, advocates, and Knowledge Holders on how the Conference of Parties of the United Nations Framework Convention on Climate Change has evolved and where it needs to go.

Inuit are one Indigenous People, bound by a shared heritage and profound connection to the Arctic landscape, yet they have been divided by arbitrary colonial borders that fragment their homeland. This chapter underscores Inuit contributions to the United Nations Framework Convention on Climate Change (UNFCCC) Conference of the Parties (COP), as we approach the thirtieth anniversary of this critical international dialogue. The contributors to this chapter are current and former COP delegates, spanning from what is now known as Canada, Alaska, and Greenland. The co-authors come together with a unified voice, defying state borders and highlighting their shared struggle for climate justice. While Chukotkan Inuit are full members of ICC, there has not been participation in the UNFCCC COPs from ICC Chukotka in recent years. Nevertheless, ICC Chukotka contributes to formulating ICC's position papers, enabling Inuit delegates to advocate with a unified and truly representative voice at international meetings.

For decades, Inuit have been on the frontlines of climate change. Indigenous Knowledge is intimately intertwined with the Arctic ice and land, offering a unique and invaluable perspective on the transformations occurring in the polar regions.[1] As temperatures rise and sea ice retreats,

---

1. ICC defines Indigenous Knowledge as: "Indigenous Knowledge is a systematic way of thinking applied to phenomena across biological, physical, cultural, and spiritual systems. It includes insights based on evidence and acquired through direct and longterm experiences and extensive and multigenerational observation, lessons, and skills. It has

Inuit experience firsthand the profound disruptions to their ways of life. As noted by the Intergovernmental Panel on Climate Change (IPCC), "the polar regions are losing ice, and their oceans are changing rapidly. The consequences of this polar transition extend to the whole planet."[2] From changing hunting patterns to shifting ice conditions, the impacts of climate change are not abstract or "economic and non-economic"—they are immediate and, in many cases, past the tipping point of adaptation. As Elder Piita Irniq notes, "We are experiencing major changes that we have not seen before, not in our lifetime or in the past. We have lived with climate in peace and harmony for thousands and thousands of years. Since the 1960s and 70s we started to see some changes, we started to notice weather changes that were impacting our people in the Arctic in a big way."

This chapter illuminates how Inuit have consistently raised their voices on international platforms, advocating for climate action that respects their rights and acknowledges their experiences. Their messages have resonated through numerous COPs, emphasizing not only the urgency of climate action, but also the need for the world to recognize and act on the profound interconnections between climate change, the Inuit cultural context and knowledge, the integrity of the Arctic environment, and human rights of Indigenous Peoples.

---

developed over millennia and is still developing in a living process, including knowledge acquired today and in the future, and it is passed on from generation to generation." Inuit Circumpolar Council, *Circumpolar Inuit Protocols for Equitable and Ethical Engagement* (ICC, 2022), 15, https://www.inuitcircumpolar.com/wp-content/uploads/EEE-Protocols-LR-WEB.pdf.

2.   Meredith Sommerkorn, Martin Sommerkorn, Sandra Cossetta, et. al., "Polar Regions," in *IPCC Special Report on the Ocean and Cryosphere in a Changing Climate*, eds. Hans-Otto Pörtner, Debra C. Roberts, Valérie Masson-Delmotte, et al. (Cambridge University Press, 2022), 203–320, https://doi.org/10.1017/9781009157964.005.

The participation of Indigenous Peoples and specifically Inuit in these high-stakes, often bureaucratic forums is not just symbolic—it is essential. The value of their presence cannot be overstated. Engaging in these complex and sometimes inaccessible spaces allows Indigenous Peoples' representatives to directly influence policy, challenge inequities, and drive meaningful change. Their active involvement helps ensure that climate solutions are not only scientifically sound, but also culturally relevant, equitable, and just. While the change is slow moving, it can be seen over time. Elder Alexina Kublu states, "I would like to see more Indigenous Peoples speaking in the plenary. Everyone needs to hear Inuit. People need to be educated from Inuit—they need to learn more from Indigenous Peoples and I would like to see more Inuit speaking at the highest level so that we can educate and inform the international community."

The Inuit relationship with the land, coastal seas, marine mammals, and ice is not merely a cultural connection, but a way of life intimately tied to the environment and the oceans that have sustained Inuit since time immemorial. Indigenous Knowledge, always evolving and passed down through generations, encapsulates a deep understanding of environmental changes and a thriving culture. This wisdom stands in stark contrast to the short-sighted approaches often seen in global climate discussions that focus on economic prosperity and perpetuating a state of consumption that cannot be sustained. This can be seen, for example, in the current discussions around economic and non-economic loss and damage. Inuit leaders have consistently called for policies that honour their knowledge systems and recognition of Inuit leadership in climate resilience and adaptation, not only focused on compensation for things such as losses and damages, but looking at aversion and protection for future generations.

Looking to the future, this chapter also explores the pressing need for acknowledgement and respect for Inuit self-determination in all international forums. The ability of Inuit to shape their own futures, to have their voices heard in decision-making processes, and to exercise control over their lands and resources is critical not only for their cultural survival, but for the broader fight against climate change. As the international community prepares for the next decade of COPs, this chapter argues that recognizing Inuit sovereignty and incorporating their perspectives and rights into global climate strategies are not just a matter of justice, but a necessity for effective climate action as the world struggles to define concepts such as a "just transition." As ICC Chair Sara Olsvig says, "How can a transition be just if it is not founded on human rights? While human rights are something internationally recognized, they still need to be fought for, to be a part of how climate action moves forward. Climate action cannot come at the expense of human rights."

As we stand on the brink of a new chapter in the climate change dialogue, the contributions of Inuit serve as a powerful reminder of the need for unity, respect, and action in the face of a rapidly changing world. ICC Chair Sara Olsvig continues, "The full and effective participation of all Indigenous Peoples in these forums is a crucial step towards ensuring the voices of those most affected by climate change are not only heard but are integral to crafting solutions that are both just and effective over the next 30 years and beyond." Elder Alexina Kublu asserts, "We want to make sure that we have enough wildlife for our grandchildren as well as their children in the future. A clean environment. It is our human right."

9.2. Inuit Action Over the Last 30 Years

As cultural, environmental, and human rights advocate, author and former ICC Chair Sheila Watt-Cloutier says, "We can't think our way through [climate change] anymore, we have to feel our way through. It really is about changing the hearts of people and realizing that we're all in this together."

While the UNFCCC serves as an important forum for global climate discussions, its framework is not inherently grounded in a rights-based approach. ICC Chair Sara Olsvig identifies that, "The rights of Indigenous Peoples need to be fought for time and again. Decisions made within the UNFCCC do not automatically ensure effective implementation, especially regarding the rights of Indigenous Peoples, as Inuit experiences have shown."

Indigenous Knowledge, land rights, and unique ways of life are often sidelined in broader climate agreements or added together with stakeholder groups, leaving Inuit and other rights holders vulnerable to the very environmental changes these discussions aim to address, and sometimes even the solutions pose a threat if the rights of Indigenous Peoples are not recognized. Despite the increasing recognition of the leadership role of Indigenous Peoples in climate change solutions by some, Inuit and Indigenous Peoples must continue to show up at these forums, insisting that rights be fully respected, recognized, integrated, and upheld. Often, changing political landscapes drastically impact how governments choose to uphold the rights of Indigenous Peoples and this is one reason international climate agreements must include Indigenous Peoples as rights holders and decision-makers. Without consistent advocacy and active participation, policies formed in global spaces can remain disconnected from the lived realities of Indigenous Peoples on the ground. Iñupiaq advocate and former ICC Chair Dalee Sambo Dorough is clear: "Don't tell us what the rules are, go away and figure out how to change them so that we can get something done." Only by remaining present and vocal can Inuit continue to demand change. Inuit presence has been significant. "Even though we are few in number, there is still significant impact," as Dalee Sambo Dorough continues, "I think what made the COP 24, 25, and 26 successful for Inuit is the fact that we weren't just waiting for the next negotiation. Autonomously we dealt with State party members and negotiators before, during and even after we got home—to ensure that they are going to be strong in their positioning going into the next negotiations. We need to hold them accountable for their respective commitments and their human rights obligations."

Moreover, the predictions and warnings articulated by Inuit leaders about climate change have proven prophetic. Now, as the Arctic continues to warm nearly four times faster than the rest of the planet, the world is taking note that Inuit have been bringing this message since before the Rio Earth Summit of 1992. And, as climate change progressed, Inuit continued to bring these messages, forcing the international community to hear them. Inuit have long observed the consequences of a warming planet, predating the UNFCCC, as it was and continues to be a lived reality impacting hunting, fishing, harvesting, and other lifeways of Inuit every single day. This knowledge of the environment and the changes that have been observed are not confined to the Arctic; it resonates globally, as Indigenous Peoples' insights about climate impacts and adaptation strategies are increasingly recognized in international forums. The careful balance and role of the Arctic and Antarctic cryosphere for the stability of global atmospheric, ocean, and terrestrial systems are understood by many Inuit. Unfortunately, the disruption and interference with these systems is what is being evidenced in diverse regions with little understanding of the interrelationship between them. Just as Inuit said, what happened in the Arctic did not stay in the Arctic. Despite the decade-old Inuit message, Inuit persist against all odds, continuing to ensure that their voices are heard in scientific assessments, research, and policy, including the Arctic Council's Arctic Climate Impact Assessment (ACIA), the Intergovernmental Panel on Climate Change's assessment reports, and other domestic, international, and circumpolar publications. Their experiences serve as crucial warnings, illustrating the indivisible nature of climate impacts across the globe and underscoring the need for a collaborative approach to responses that transcend borders and political agendas.

To enhance the effectiveness of Inuit participation, it is vital to establish and maintain relationships with UN member states outside of COP meetings. Many government representatives arrive at COPs with predetermined positions, often neglecting the rich, community-based knowledge that Indigenous Peoples hold and are willing to offer. By engaging in dialogue and building relationships outside of formal negotiations, Inuit leaders can foster understanding and build alliances that promote their priorities and perspectives by working with other Indigenous Peoples and finding areas of alignment with governments. While Canada and Kalaallit Nunaat (Greenland), for example, work closely with Inuit in the lead up to and throughout UNFCCC meetings, it was not always like this, and it is not always guaranteed. These relationships are essential for creating an environment where Indigenous Knowledge is not only welcomed, but integral to policy development. Furthermore, Inuit are continuing to look outside the UNFCCC for ways to influence parties, through the courts, for example. Former ICC Chair Sheila Watt-Cloutier spoke of people around the world taking states to court for their inaction on climate change. One example of this is the precedent-setting 2005 legal petition where 62 Inuit hunters and Elders from Alaska and Canada submitted a petition to the Inter-American Commission on Human Rights seeking relief from violations resulting from global warming. Sheila Watt-Cloutier notes, "All of these movements are going to be the way that real change happens. People around the world are having their lives and livelihoods disrupted as a result of climate change. The legal system is a way that states can be held accountable and forced into action."

Among all the challenges, it is also important to reflect on how far the Indigenous Peoples' movement in this space has come. Now recognized as an official UNFCCC constituency, a group with diverse but broadly clustered interests or perspectives, Indigenous Peoples at one point were not even able to take the floor in negotiations, let alone sit equally with states on a constituted body, such as the Facilitative Working Group established by the Local Communities and Indigenous Peoples Platform (LCIPP). Inuit, who are members of the sociocultural Arctic Region within the UN system, were the first to serve on this body, and directly influence its Work Plan by delivering a substantive series of training to UN member state representatives on the contours of Indigenous Knowledge. It was crucial to demonstrate that Inuit and other Indigenous Peoples are the experts on Indigenous Knowledge rather than state government representatives, Western scientists, or non-governmental organizations.

As such, participation in UNFCCC negotiations is not merely about representation; it is about transforming the climate dialogue to be inclusive and holistic through whatever means or areas of influence possible. By enhancing participation wherever possible, de-siloing the discussion, shifting emotional perspectives, and building relationships with states, we can work towards a future where Indigenous Knowledge and worldviews are at the forefront of climate action. Recognizing the truths that Inuit have shared—and the dire consequences of ignoring them—will be vital in shaping a sustainable and equitable response to the global climate crisis.

9.3. Where We Are

The landscape of UNFCCC negotiations presents a daunting challenge for Inuit and other Indigenous Peoples, who often find themselves navigating a state-driven space dominated by non-governmental organizations (NGOs) and other entities with vast resources and established influence. These organizations frequently possess extensive budgets, enabling them to engage in high-profile advocacy and outreach that, for too long, have overshadowed the voices of Indigenous

Peoples. Despite this imbalance, the perspectives and experiences of Inuit have been influential and crucial for informing and enriching the discourse. As the current ICC Chair Sara Olsvig notes, "I want to make it very clear that both in terms of the UNFCCC in particular and the UN in general, Indigenous Peoples are contributors of invaluable and indispensable perspectives and ways of working. I have an ambition that we become even more active in the actual political negotiations."

While participation in these forums has grown in recent years through forums such as the LCIPP, there has also been a recent back step in terms of transparency in the process, a growing concern for Indigenous Peoples who work tirelessly on text to ensure there are rights safeguards. We have seen these safeguards removed without explanation, for instance at COP29 where language that Indigenous Peoples had been fighting for was dropped out of the agreement on the New Collective Quantified Goal on Climate Finance. Indigenous Peoples' contributions are indispensable to the negotiations. We saw this occur at COP29 where language that Indigenous Peoples had been fighting for was would be dropped out of the agreement on the New Collective Quantified Goal on Climate Finance. Indigenous Peoples' contributions are indispensable to the negotiations. They offer unique insights into adaptation and mitigation strategies that have proven effective in their own communities, showcasing resilience and innovation in the face of climate change. This wealth of knowledge provides a more comprehensive understanding of climate impacts, enabling negotiators to consider solutions that incorporate diverse ways of knowing and being in the world. Indigenous Peoples continue to fight for the recognition of rights and Indigenous Knowledge within agreements and, step by step, line by line, the fight will pay off.

Historically, Indigenous Peoples, including Inuit, faced significant barriers in these discussions. For many years, they were unable to speak in plenary sessions, effectively silencing their voices in critical conversations about issues that directly affected their lives and territories. While Canada and Kalaallit Nunaat have emerged as more supportive allies in recent years, this shift has not always been the case. While it should be recognized as a legal obligation of Canada to meaningfully engage with First Nations, Inuit, and Métis, it is rather seen more as a showing of good will by the government of the time. The evolution of the country's stance reflects a broader recognition of Indigenous rights and Knowledge, but there remains a legacy of exclusion that continues to influence the dynamics of these negotiations. "It's a big job to speak for our people, when we are up against country interests, industry interests, even NGO interests," said former ICC Vice-Chair Lisa Koperqualuk. "We are dealing with some countries that have absolutely no recognition of Indigenous rights, so I do have to say that Canada has become a model in recent years." This relationship and this work, for both Indigenous Peoples and party representatives, needs to be nurtured and needs to transcend political boundaries and timelines—climate change will not wait. Indigenous Peoples need a permanent voice at the table that does not align them with environmental organizations and "stakeholders." Indigenous Peoples are right holders!

Indigenous Peoples bring invaluable recommendations, knowledge, and lived experiences that can significantly enrich the UNFCCC dialogue. Their holistic understanding of and interconnected relation with the environment—rooted in centuries of stewardship and cultural practices—offer perspectives that are often lacking in conventional climate discussions, which may focus narrowly on economic or technical solutions. The disconnect between the implementation of the United Nations Declaration on the Rights of Indigenous Peoples (UNDRIP) and its recognition in practice must be acknowledged and addressed. Many Indigenous Peoples still face systemic barriers that prevent their meaningful participation in decision-making processes, undermining the very principles of UNDRIP. ICC Chair Sara Olsvig discusses some of these challenges:

> Other processes do not automatically trickle down into other UN agencies and negotiations. If we make progress at the Human Rights Council, for example, there is a recurring challenge to implement the rights that we've fought for in other forums and in related texts in climate change negotiations or other UN bodies. That's a big challenge, but as we've seen with the Paris Agreement, Indigenous Peoples and allies were able to push for a rights-based approach and see that reflected in the final document, which is a huge accomplishment. The UN as an institution is siloed, creating barriers for Indigenous Peoples and their rights to be reflected. Nonetheless, Inuit have succeeded in putting climate change on the agenda and we should recognize the successes that past leaders have fought decades for.

As discussed, one significant challenge within the UNFCCC is the tendency to operate in silos, where different sectors and issues are treated separately. One clear example is the matter of human rights, often dismissed because it is being discussed and addressed elsewhere within the UN system. To address climate change effectively, we must de-silo the conversation and recognize the interconnectedness of political, economic, environmental, social, and cultural issues. Indigenous Knowledge and practices, honed over millennia, can offer transformative insights that bridge these divides. Inuit experiences highlight the urgent need to shift from merely thinking about climate change to feeling its impacts in a visceral way, prompting a change of heart that drives collective action. By engaging emotionally with the realities faced by Indigenous Peoples, we can cultivate a deeper understanding of the urgency of climate action for everyone.

The challenges faced by Inuit and other Indigenous Peoples within the UNFCCC negotiations highlight the urgent need for systemic change. As allies, governments like

Canada must actively support and facilitate the inclusion of Indigenous Peoples, recognizing their essential role in shaping climate policy, not only in a tokenistic way, but in incorporating their red lines as their own. By valuing Indigenous Knowledge and promoting genuine, comprehensive, full, effective, and meaningful participation, we can move towards a more holistic and effective approach to climate action that honours the rights and experiences of all.

9.4. The Future of Inuit Advocacy at Climate Change COPs: Youth Perspectives

As former Canadian National Inuit Youth Council (NIYC) President Susie-Ann Kudluk says, "My hope for the future is that our land is intact, that they get to experience hunting the same animals that we have today. My message for Inuit continuing this fight is to know that your ancestors are there with you, they are there to guide you. Just believe in yourself and speak up when you feel the need to."

Inuit youth understand that the health of the planet is intrinsically linked to their well-being; when the land, sea, and ice thrive, so do they. Their priorities are not rooted in monetary wealth or capitalism; instead, they value the health of their communities, the resilience of their ecosystems, and the rich cultural heritage that is the foundation of their culture and relationships.

Increased access to training in these forums is needed and Inuit youth recognize this as urgently needed across all platforms. As Inuk youth advocate Miyuki Daorana outlines, "We can avoid climate injustice and broken promises through training everyone in platforms with genuine relationship building, ethics, empathy, and especially justice." Inuit youth assert that climate change discussions should not solely revolve around numbers and economic forecasts; they must be infused with diverse knowledge systems, particularly those of Indigenous Peoples. Strong trust and collaboration are needed to build a true and equal foundation because the norm of formality and professionality in a lot of these spaces limits the holistic and realistic context. This is essential for fostering a truly inclusive forum that deviates from the current state-centred approach to climate negotiations in order to effectively change the trajectory for the future.

In navigating the complexities of climate change, Inuit youth emphasize the importance of considering how current actions will affect future generations. They pose critical questions about how those generations will experience the cultural activities that allow them to be connected—gathering berries, hunting on the land, or travelling on the ice. The decisions made in these negotiations will ultimately determine whether their children can inherit the rich traditions and natural beauty that define their way of life. They advocate for solutions that not only address the present crisis but also pave the way for a sustainable and thriving future.

For instance, Susie-Ann Kudluk shares, "The ways that climate change impacts me is that the land is very different from when I was little due to rapid rates of climate change. The seasons are going by a lot faster, the migration routes of animals that we hunt have changed significantly, causing us to adapt to unfamiliar territory. This is a direct threat to our culture and to future generations."

At COP, many Inuit youth felt empowered to take up space, for example at the LCIPP roundtables, where parties listen and respond to statements made by youth. The platform provided an opportunity for them to network, express their concerns, and share their visions for the future. Meeting other youth from around the world ignited a sense of solidarity and purpose, leading to the formation of a network dedicated to action. As they engage in these critical conversations, Inuit youth called for the UNFCCC to improve its approach to knowledge sharing across different systems. They stressed that integrating Indigenous Knowledge into climate policy is not just beneficial; it is essential. This integration provides a holistic understanding of environmental challenges and solutions that mainstream approaches often overlook. As Miyuki Daorana continues, "Our love for earth is ancestral and beyond ourselves, because the health of our planet is the health of us […]. Economic wealth and capitalism is not our highest priority and this is the fundamental shift that is needed in these spaces."

The commitment of Inuit youth to the Earth is a legacy that transcends time. They stand united, ready to advocate for a future where the health of their planet and their people is prioritized, where ethical leadership guides decision-making, and where every voice, especially those of young people, is recognized and valued. Together, they strive to ensure that future generations inherit a world where they can continue to live in harmony with nature, just as they have been able to do. They continue to leave their families and their homes to advocate for Inuit and for future generations, entering spaces that are all too often unfamiliar and unsafe.

Miyuki Daorana is clear, "I hope for a just and equal international platform with an ethical mindset and ethical decision making. A place where ethics and empathy are placed higher than capitalism. The UN must do better. For all of us and for future generations."

9.5. Moving Forward

Inuit must continue to participate actively and take up space in climate negotiations, applying pressure in all possible areas to drive meaningful change. While the current frameworks often do not cater to Inuit, their ongoing presence is crucial; without it, the status quo will persist. Indigenous voices need to be elevated in these negotiations, ensuring that their perspectives are not only included, but prioritized. Furthermore, equitable access to climate finance resources is essential for enabling Indigenous Peoples in all regions to engage fully in these mechanisms, empowering them to advocate for solutions that honour Indigenous Knowledge and experiences. By remaining active participants, Inuit can help reshape the narrative and influence outcomes that will benefit their communities and all future generations relying on us.

> My hope for the future is that my children, grandchildren, and great-grandchildren get to experience the amazing life I got to experience when I was a little girl, out on the land, on the tundra, on the sea.
>
> —*Susie-Ann Kudluk*,
> former NIYC President

CHAPTER 10

Otipemisiwak and Climate Leadership: The Métis Nation at UNFCCC

Dane de Souza and Kate Gillis

The first time I participated in the United Nations Framework Convention on Climate Change (UNFCCC), or any UN space for that matter, was COP27 in Sharm el-Sheikh, Egypt. This was the first time ever the Métis National Council (MNC) supported and sent a full delegation, made up of representatives from across the Métis Homeland, to a UNFCCC COP. After years of effort, advocacy, and capacity building, the MNC was able to develop the resources to overcome financial and logistical barriers to attending COP. I also had the great privilege of co-leading our delegation in various events related to Indigenous Climate Leadership, engaging in and supporting Canada on negotiation items, and taking part in the closing ceremony of the Canada Pavilion. Events ranging from panels on water protection to Indigenous Science and Knowledge, as climate solutions at the Canada Pavilion provided a platform in which the MNC delegation could share solutions and further conversations with a global audience.

Having started my career path as a wildland firefighter in rural Alberta to finance my undergraduate studies in international commerce, to chasing my passion for wildfire during my masters of international forestry, to researching the Métis connection to Indigenous fire stewardship and working for the Métis Nation, the UN always seemed like too exclusive and complex of a place to find my way to. When I first walked through the gates, several things went through my head:

1. I just got here, and I'm completely exhausted from logistics, travel, learning the technical language of the UNFCCC (a brand of English foreign to even native speakers) and wrapping my head around COP […] and we haven't even begun.
2. The younger version of myself screaming, "NO WAY! We're at the UN?! How did we go from growing up in Calgary to this? What a dream come true!"
3. All of those who came before me, all the Métis folks, all my Indigenous kin who fought for my ability to be here today sacrificed more than I'll ever know for this opportunity. Time to go to work for them and for those who come next.

The UNFCCC spaces are exhausting. The physical, mental, and emotional toll of these spaces is immense: long days of using all one's mental capacity to navigate the complex language, negotiations, and inner workings of the UN, exacerbated by subpar nutrition (should you have time to eat) and disrupted sleep schedules, and listening to stories of communities and lives disrupted by climate impacts and seeing one's community reflected in those impacts. It is a challenge that reminds me of working long shifts on wildfires during my time as a wildland firefighter. However, for Indigenous Peoples, particularly, the impacts mentioned above are further heightened and carry a unique component of spiritual exhaustion. As Indigenous Peoples in this space, we do not get a vote, and our voices are often actively suppressed. As Indigenous Peoples, we do not get to vote or directly take part in negotiations in the same manner that nation-states like Canada or Luxembourg do. Instead, Indigenous Peoples from every corner of the globe must combine our efforts to lobby nations and find innovative ways to have our voices heard.

Yet our connections to our lands, waters, skies, and all those who dwell within them are directly and acutely impacted by the very negotiations and conversations we find ourselves sidelined for.

When I get home from these spaces and talk to my Métis community, my family, and my friends, I am often asked, "So what are you guys doing over there?"

First, our participation in these spaces is an act of sovereignty. An opportunity to stand alongside our Indigenous kin from across Mother Earth and work, teach, learn, and contribute to the betterment of the communities we belong to. From the Sami of Lappland (Finland, Sweden, and Norway), to the Yanomami of the Amazon Rainforest (Brazil), to the Māori of Aotearoa (New Zealand), the Indigenous Peoples Caucus collaborates to forward Indigenous climate leadership and sovereignty via lobbying, position statements, and hosting events. Walking through the doors of the UN alone is an act of sovereignty and a statement that the MNC internationally represents the Métis Nation wherever our rights and identity can be protected and championed. Beyond sovereignty, participating in these spaces alongside our Indigenous kin honours the Métis tradition of stewardship. A core component of Métis culture, identity, and spirituality is our reciprocal obligations to the lands, waters, skies, plants and animals that sustain us. As Métis people, we have an obligation to steward and protect these relationships, whether it be through adhering to protocols around harvesting or taking part in UNFCCC spaces that will impact how climate change affects those relationships.

At a more technical level, the MNC's work within the UNFCCC is an opportunity to build relationships with other nations, Indigenous Peoples, NGOs, researchers, private industry, and others working to combat the climate crisis. Our time at COP27 and COP28 and the interim pre-sessional meetings have provided the opportunity to build a strong relationship with Canada's UNFCCC negotiators. As much as we are elated and grateful to build these relationships and work together, this is also a space in which we, as a nation, can hold Canada accountable. The realities that Métis citizens in Canada experience, because of colonial policy and structures, is not always reflective of how Canada represents itself in international forums like the UNFCCC. By being at these tables, by being in these rooms, and by being able to connect our nation to them, we can ensure that the climate realities faced by Métis citizens do not go misconstrued or unheard. It is not enough that we highlight the impacts of climate-related disasters like drought, floods, and wildfires have on our communities, but also highlighting the climate solutions held within Indigenous Knowledge and Science that steward 80% of the world's biodiversity through Indigenous land stewardship techniques like Indigenous fire stewardship and Indigenous agricultural techniques.

A component of my time within UNFCCC that I will always cherish is the Indigenous solidarity and kinship experienced through the Local Communities and Indigenous Peoples Platform (LCIPP) and the International Indigenous Peoples Forum on Climate Change (IIPFCC). Walking into that room filled with folks from the Amazon Rainforest to Lappland, and everywhere in between, inspires hope and motivates sustained action. Seeing my Indigenous kin from across the globe proudly representing their Indigeneity is a testament to the resilience of our cultures, identities, and nations. When juxtaposed with the sea of blue suits at the UNFCCC, it is a stark example of the humanity that needs to be upheld, promoted, and protected at the negotiation tables.

To sit with a Maasai leader, a Māori researcher, an Ojibwe Elder, and a Knowledge Keeper, and build friendships, discuss the lands to which we belong, and collaborate to protect that which we are all so closely connected too […] it gives more hope than anything else I have seen at the UNFCCC. In this space you have individuals from diverse cultures, environments, and walks of life, who speak different languages and are all so radically different, and yet we are all united and collaborating for the collective good of all Indigenous Peoples, Mother Earth, and all who dwell on her. In one moment, you can be sharing a laugh with a colleague and friend you have not seen since the last UNFCCC meeting, and in the next it is straight down to business to strategize and work to advance Indigenous climate leadership. If anything, the LCIPP has taught me that being "Indigenous" does not only mean that we as Indigenous Peoples share a colonial history, but we share a profound and sacred connection to the lands, skies, and waters we belong to. Current colonial powers and systems, like the UN, may try to paint Indigenous Peoples as victims and nothing more; however, my colleagues within the LCIPP have consistently shown that we are stewards and champions of what the UNFCCC is trying to achieve: We are resiliency, we are adaptation, we are mitigation, and we are sustainability. Our Indigenous communities have been and continue to be resilient, find adaptation solutions, and mitigate circumstances, which have sustained us through ecological, economic, social, and colonial changes; and our knowledge, our science, and our cultures hold lessons for all humanity to learn from, particularly in the context of climate change.

From the first time I walked onto the UNFCCC grounds in Sharm el-Sheikh, I have kept the mindset of creating building blocks for those who come next. Those who come next could be here as soon as tomorrow to take this work further than I can, so my time in these spaces is never guaranteed. Each day is the last day. The UNFCCC is not a place where climate change gets "solved." That happens at the landscape level, where decisions are made every day on how we choose to steward or extract from nature. The UNFCCC is a space in which people from every walk of life can come to offer the solutions, information, and realities around the single greatest threat to humanity's existence. Although this space is profoundly exhausting, although the bureaucracy and political posturing are sickening, I still leave these spaces with hope. Hope inspired by the people who will do the work, the people who will solve these problems, not the governments nor the institutions that perpetuate them for short-sighted, self-interested gains.

Finance is not a solution. Carbon markets are not a solution. Just transition is not a solution. Loss and damage funds are not a solution. These mechanisms may be part of the solution so long as they support Indigenous-led, landscape level, environmental stewardship activities that bolster biodiversity, carbon sequestration, and ecological resilience. These are the solutions that have sustained humanity since time immemorial. "Decolonization" has become a buzzword used by colonial institutions like the UN and the nations signed to the United Nations Declaration on the Rights of Indigenous Peoples to pay lip service to recognizing the humanity of Indigenous Peoples. Actioning Indigenous climate leadership and the solutions within it humanizes Indigenous Knowledge and Science as landscape-based actions that benefit all of humanity regardless of borders and ideologies.

In recent years, COP has become a fairground for career advancement and attaining celebrity status. So often folks will expend the great amount of resources and time required to attend COP with the ultimate goal of becoming the next Greta Thunberg or Leonardo DiCaprio. Although the intention is ultimately geared towards being part of finding and mobilizing climate solutions, the media attention on events like COP has created a common mistrust and lack of faith in the ability of the UNFCCC to provide the solutions it is tasked with. For anyone with the ambitions of attending COP to have their voices heard or help to find these vital solutions, my advice is to approach this space with humility and an open mind, heart, and set of ears. Do not allow yourselves to be discouraged by the apathy and false solutions that dominate negotiation outcomes. Instead, use this time and space to find your place in forwarding landscape-level solutions that provide economic, social, and cultural opportunities to sequester carbon and realize sustainable interactions with the lands, skies, waters, and life forms that sustain human life on planet Earth.

The United Nations Framework Convention on Climate Change (UNFCCC) Conference of Parties (COP) is a paradigm of opportunity and sacrifice, which can be examined broadly using Bronfenbrenner's ecological systems theory. Examining the interconnected systems, ranging from the immediate surroundings to broad societal structures, the COP is a space of competing realities, structures, policies, and beliefs, which have the potential to impact international, national, regional, local, and individual rights and interests. Or at least, that is the way it has been sold to us.

As Indigenous Peoples navigating international spaces, we are constantly reminded that we are participating in a system that was not built for us. Compound that with any additional "equity" groups, including being a woman, being a youth, being a part of the LGBTQ2S+ community, and so on, the United Nations is an inherently hostile space. It is important to be mindful of the history of the United Nations as a system, and how, at its core, is a need to maintain institutional hierarchy. And yet, Indigenous Peoples have adamantly fought for their seat at the table since 1923 when Haudenosaunee Chief Deskaheh petitioned the League of Nations for recognition of nationhood. In the 100 years that have followed, Indigenous Peoples have made significant strides, including the creation of the Local Communities and Indigenous Peoples Platform (LCIPP) within the UNFCCC system.

COP30 will be my third UNFCCC COP. When I was still in my MA program, learning about and trying to get involved in multilateral international spaces, to me, attending COP was a sign that you made it. I remember watching the outcomes of COP26 through the media, including Txai Suruí, a youth climate activist from the Brazilian Amazon, telling world leaders that, "Indigenous Peoples are on the frontline of the climate emergency, and we must be at the centre of the decisions that are happening here," and thinking, "I need to be there, that's where change is happening." Two years later, that dream came true with the Métis National Council delegation, and I had the chance to attend COP28 in Dubai, and then COP29 in Baku.

It is interesting to witness the conversations on the ground, as opposed to what is shown in the media. Conversations among the Indigenous Caucus oftentimes highlight the dichotomy I raised above of sacrifice and opportunity. Indigenous People are often questioning whether being at these big United Nations forums is actually worth it, or if they would be better off back home in their communities working on grassroots solutions. And yet, they keep coming back.

As a Métis woman and youth in this space, I ground myself in the worldviews that I was raised in *Otipemisiwak* (Cree term used to describe the Métis as "the people who own themselves) and *Wahkohtowin* (Cree term roughly translated to "kinship" or "we are all related"). We show up in these spaces because it is our inherent right, recognized under the United Nations Declaration on the Rights of Indigenous Peoples, to do so. We also show up in obligation to all our relations: our communities back home; our Indigenous kin around the world; our Mother Earth, and her lands, waters, plants, and animals; and all our generations, both past and present, to ensure that this self-determination is both protected and promoted.

We do this in a way that acknowledges and embraces opportunity and sacrifice as being mutually inclusive and reinforcing as an avenue for progress. We do this in a way that looks past micro, meso, exo, macro, and chronosystems, and again reiterate these systems as being mutually inclusive and reinforcing. Because at our core, we are taught *Wahkohtowin*, that everything is related. The COP would benefit from acknowledging these worldviews, self-determination, and kinship in all decision-making processes. Until then, Indigenous Peoples remain ready and willing to be recognized as equal partners to restore our relationships with the planet, and with one another.

If you are considering attending COP30 and beyond, I encourage you all to question why. I have firsthand witnessed the time, energy, capacity, and resources that go into the UNFCCC trying to find solutions to climate change, without anyone actually considering the root cause, which is colonialism. And when we think about colonialism, we also have to think about the intrinsic connection to capitalism and patriarchy and all the systems and structures that continue to uphold these constructs. Indigenous Peoples have long understood that environmental health is inseparable from community well-being, cultural survival, economic autonomy, and so on. Our knowledge systems emphasize balance—not just between people and the land, but between generations. When considering solutions, it is critical to think both past and present, how we got here and where we are going.

COP is a critical space for advancing the conversations on climate change, but these conversations cannot happen in silos and must work towards addressing the root cause—until then, any solutions continue to be false.

CHAPTER 11

The UNFCCC COP:
An Imperfect—and Essential—
Civic Space for Climate Multilateralism

Caroline Brouillette

E very year around the end of November, it is the same refrain. From all corners of the political spectrum and media commentary, and including within the climate movement itself, the annual climate Conference of the Parties (COP) is criticized, ridiculed, and called into question.

Some of the issues raised are well worth pondering. For instance, was it really necessary to mobilize 644 private flights and the associated greenhouse gas emissions for the Dubai COP? How are we still allowing fossil fuel lobbyists—numbered at more than 2,400 for COP28 and nearly 1,800 for COP29—to participate in a process that aims at limiting the harms caused by the product they sell?[1] How do the pavilion space and Green Zone roadshow, with tens of thousands of attendees, really contribute to multilateral negotiations?[2]

---

1. Kick Big Polluters Out, "Release: Record Number Fossil Fuel Lobbyists Attend COP 28," news release, December 5, 2023, https://kickbig polluters out.org/articles/release-record-number-fossil-fuel-lobbyists-attend-cop28.

2. COP spaces are divided into a Blue Zone and a Green Zone. The Blue Zone is the formal, UN-managed space for official negotiations and is

Why did it take 30 years before naming the combustion of fossil fuels—the main culprits of climate change—in United Nations Framework Convention on Climate Change (UNFCCC) decision texts?

It is also necessary to ask what these conferences have achieved, as the evidence points to accelerating climate change. In 2024, the world reached the symbolic threshold of 1.5°C. While the Paris Agreement limit is a long-term target of 20-plus years and thus has not been crossed yet, we are seeing and feeling the impacts of the crisis and unnatural disasters caused by the burning of fossil fuels multiplying in Canada and across the world. At the time of writing, the World Meteorological Organization had endeavoured to tally the number of climate disasters that have unfolded over the past year and ended with the heartbreaking number of over 150.[3] With record-breaking wildfires, floods, and heatwaves hitting homes and devastating lives and communities, it is valid to wonder if and how COPs are actually helping.

Some of the criticism is more questionable. I remember when, a few years ago, some journalists started digging into the size of the organization I lead, Climate Action Network Canada (CAN-Rac)'s COP delegation, and whether we had calculated our carbon emissions and used carbon credits to compensate for the climate impact of our flights. They were, of course, uninterested in hearing about the advocacy work we did on the ground and the emissions that might have been avoided as a result of this work.

But what if more than anything, and despite the pressing need for reform, COPs—and civil society's essential role as observers and rights-holders—were mostly misunderstood?

---

restricted to accredited delegates, while the Green Zone, managed by the host country, is more open to the public, businesses, and civil society.

3. Damian Carrington, "More Than 150 'Unprecedented' Climate Disasters Struck World in 2024, Says UN," *The Guardian*, March 19, 2025, https://www.theguardian.com/environment/2025/mar/19/unprecedented-climate-disasters-extreme-weather-un-report.

***

I became a climate activist in 2018—the year scientists came out with a shattering report warning of the consequences of global temperature rise above 1.5°C. I was 26 years old, and as a figure skater who grew up spending the winters practising at the outdoor rink at the park near my home in Sherbrooke, I was witnessing how winters were shortening. The game changing Intergovernmental Panel on Climate Change report spelled out how the next 12 years would be critical for bending the global emissions curve.

That was only three years after the signature of the landmark Paris Agreement, which called on parties to "limit global warming to well below 2°C, preferably to 1.5°C, above pre-industrial levels." While we are starting to approach these dangerous thresholds, we need to consider a pressing counterfactual: What would have happened *without* the Paris Agreement being signed?

According to the International Energy Agency, current policies and pledges would lead us to a temperature rise of 1.9°C in 2050 and 2.4°C in 2100.[4] This would by far exceed the Paris Agreement goals, and, as we are already seeing and feeling, with every tenth of a degree of temperature increase, would be catastrophic for humans, ecosystems, and the economy. Yet, before the Paris COP, global carbon emissions were on track to cause a devastating temperature rise of nearly 4°C by the end of the century. It may be concluded, then, that climate multilateralism is working, it is just not working fast enough.

The reasons for that, though, have more to do with politics than the framework itself. While many have lost their

---

4. "World Energy Outlook 2023: Secure and People-Centred Energy Transitions," International Energy Agency, accessed April 8, 2025, https://www.iea.org/reports/world-energy-outlook-2023/secure-and-people-centred-energy-transitions.

breath about the nationally determined and non-binding nature of this treaty, more attention needs to be paid to domestic politics and how that shapes the policies that countries are able to contribute to the global climate fight. Michaël Aklin, associate professor of political science and public policy at the University of Pittsburgh, and Matto Mildenberger, assistant professor of political science at the University of California, Santa Barbara, find that the assumption that climate action is a global collective action problem structured by free-riding concerns does not hold empirical water.[5] Instead, patterns of climate policymaking can be better explained by distributive conflict within countries. They find that defections from global agreements—like the United States from the Kyoto Protocol and then the Paris Agreement (again)—have not resulted in weakened commitment from other countries.

As Kaveh Guilanpour, vice president for international strategies at the Center for Climate and Energy Solutions, puts it: "Blaming the Paris Agreement for inadequacies in global climate action is akin to a bad driver blaming their car for a crash. A better car is not the answer: getting better at driving is."[6]

The Paris Agreement was built and serves as a mirror to reflect to the world what these national policies and political will amount to at a global level; to point the way to what is needed to achieve its goals; and to generate political peer pressure to drive accelerated action.[7] Part of the world's COP

5. Michaël Aklin and Matto Mildenberger, "Prisoners of the Wrong Dilemma: Why Distributive Conflict, Not Collective Action, Characterizes the Politics of Climate Change," *Global Environmental Politics* 20, no. 4 (2020): 4–27, https://doi.org/10.1162/glep_a_00578.
6. Kaveh Guilanpour, "The Paris Agreement: A Moment for Reflection. Discussion Paper," Center for Climate and Energy Solutions, January 2025, https://www.c2es.org/wp-content/uploads/2025/03/The-Paris-Agreement-A-moment-for-reflection.pdf.
7. Kaveh Guilanpour, "The Paris Agreement: A Moment for Reflection."

cynicism may be explained by mismatched expectations. Not everything can be won (or lost) at COP, and to maximize chances of success, investing the domestic political arena is where most movement on climate action can be imposed on countries, especially in democracies. Therefore, while there is a lot that negotiations can achieve—especially when they are influenced by a smart and strategic civil society—many outcomes are already locked in once we get to the annual COP.

***

COPs have evolved from their initial purpose of negotiation between parties. A summit of world leaders, an annual rendezvous of the who's who in the climate space (and increasingly those who are very tangentially connected to it), an exhibition space for clean (and very often less than clean) technologies, an event that people go to try to get on panels are among the parts that have been attached to the convenings' main and wonky core business of negotiations.

Yet, the relevant questions about whether the now over 50,000 annual COP goers truly contribute to the process remain. Among the growing number of COP attendees, civil society's presence is absolutely fundamental, not only to ensure that a diverse set of voices are represented in COP negotiated texts, but to exert power to balance corporate interests and protect the global climate cooperation process—as imperfect as it is.

Over the years, I have heard of and witnessed many of these powerful contributions by civil society. One such example happened at COP23 in Bonn, which was presided over by Fiji. Canada's then-environment minister, Catherine McKenna, had shown up with the new and groundbreaking commitment to phase out coal-powered electricity and launch the Powering Past Coal Alliance. Yet a key component was missing: a plan for workers and communities. CAN-Rac and its labour members developed a plan to call in McKenna, who is an ally of the just transition movement. Mike de Souza of the *National Observer* was the only Canadian reporter on the ground at COP23 and covered civil society's concerns, which reverberated through Canadian mainstream media which then picked up on this reporting. Due to those advocacy efforts and coverage by Canadian media, McKenna basically agreed on the spot to set up a mechanism to propose solutions to the crucial question of how this phase-out would affect people. This led to the creation of the Task Force on a Just Transition for Canadian Coal Power Workers and Communities.[8] The task force produced two excellent reports and eventually led to the next wave of just transition policy in Canada, and the adoption of the Sustainable Jobs Act, an essential framework legislation.

Another example of civil society's capacities happened in Egypt, during COP27, where the international climate movement helped make the imprisonment of Egyptian democracy activist Alaa Abd el-Fattah an international news story. Ahead of the Sharm el-Sheikh conference, Egyptian civil society and Alaa's family had been raising concerns about his political imprisonment and hunger strikes in protest, raising fears he may die while officials attend the summit. A leader of Egypt's 2011 Arab Spring uprising, Alaa Abd el-Fattah was imprisoned by the Egyptian authorities and spent most of the past decade behind bars in a high-security prison on charges of "spreading false news" for sharing a social media post about torture. The global movement held protests inside the COP space, where Alaa's sister came in to raise awareness. Those efforts, combined with the high concentration of international media in the COP space, got the spotlight on Alaa's plight, on the issue of political repression and imprisonment

8. "Task Force: Just Transition for Canadian Coal Power Workers and Communities," Environment and Climate Change Canada, last modified March 11, 2019, https://www.canada.ca/en/environment-climate-change/services/climate-change/task-force-just-transition.html.

in Egypt, and on the inextricable links between climate justice and human rights.

In Dubai, at COP28, civil society achieved something that had never been done before and a political signal that was essential to keep 1.5°C alive: naming the need for a just transition away from oil, gas, and coal in the conference's formal decision.[9] This was the direct result of our global movement's activism, as there was no agenda item in the formal process that mandated parties to discuss this crucial issue. An issue that was made even more prominent because the COP's president that year, Dr. Sultan Al-Jaber, was the CEO of ADNOC, the United Arab Emirates' national oil and gas company. This drew fury from across the world and in much media coverage, and led Al-Jaber to want to prove that he could achieve this groundbreaking decision despite his credentials. Of course, this historic but insufficient result was the result of years of campaigning, notably from the Global Gas and Oil Network and its members.[10] A key milestone was the publication of the International Energy Agency (IEA)'s first energy production scenario aligned with 1.5°C.[11] In 2021, the IEA—whose mission is to ensure global energy market stability—had recommended that there be no new oil and gas developments or new coal mine expansions to achieve carbon neutrality, and that investments should instead be directed towards new renewable energy supply projects. This gave a huge stamp of credibility to activists and frontline land defenders who, for

9. *Outcome of the first global stocktake,* UNFCCC, 5th Sess, UN Doc FCCC/PA/CMA/2023/L.17 (2023) Dec /CMA.5, https://unfccc.int/sites/default/files/resource/cma2023_L17_adv.pdf.

10. The first Global Stocktake decision includes a historic call for accelerating action this decade to transition away from fossil fuels. Yet the text leaves the door open to carbon capture and storage, blue hydrogen, nuclear, biomass and so-called transitional fuels. Reliance on these unproven technologies are last-ditch efforts to preserve the fossil fuel industry's grip and are not in line with 1.5°C pathways.

11. https://www.iea.org/reports/net-zero-by-2050.

years, had been fighting against new fossil fuel infrastructure projects at the grassroots level. The Dubai decision added further political weight and backing to the scientific and moral imperative of a transition away from fossil fuels and scaling up of renewable energy and energy efficiency. That said, as we are seeing almost two years later, the real economy does not end at COP. The key question, as always, is about how we bring COP decisions back home for implementation and hold actors—private and public—accountable to them.

COPs are also moments when those already implementing climate action at home are able to showcase and highlight the evidence. I remember, at COP25 in Madrid, a First Nations Chief from Yukon took the floor in a CAN-Rac daily meeting to offer some words of wisdom at the end of two weeks of dispiriting and disappointing negotiations. Indigenous leaders and communities, he reminded us, are already leading the way on energy transition, as Canada's largest owners of renewable and clean energy assets after public utilities and crown corporations.[12] They are weaning their communities off diesel and gas dependency and concretely building their energy sovereignty, which is intrinsically connected to their inherent rights, and their stewardship of the land as Indigenous Peoples.

***

While these examples are some of the splashier ones, civil society has also developed some tools and coordinated movement efforts and voices in ways that have almost become part of the decor at COP. Yet despite their almost mundanity, this infrastructure has transformed the way the global climate movement works together and is able to influence negotiations.

---

12. Indigenous Clean Energy, *Accelerating Transition: Economic Impacts of Indigenous Leadership in Catalyzing the Transition to a Clean Energy Future Across Canada* (Indigenous Clean Energy, 2020), https://indigenouscleanenergy.com/wp-content/uploads/2022/06/ICE-Accelerating-Transition-Data-Report-web.pdf.

In fact, Climate Action Network International (CAN) was founded to convene civil society during COPs. Climate Action Network Canada (CAN-RAc), which I lead, is a national node of this global network, and brings together close to 200 organizations operating from coast to coast to coast. Our membership assembles environmental groups together with trade unions, First Nations; social justice, development, health, and youth organizations; faith groups; and local, grassroots initiatives. Globally, CAN—which rallies now close to 2,000 organizations from over 130 countries—was created inside COP, with the express objective of bringing together climate activists and advocates from around the world, to facilitate sharing information about their national policies and delegations' positions and coordinating strategies towards ambitious and inclusive outcomes. Over time, it has evolved into a sophisticated movement infrastructure that not only holds daily coordination meetings, which help activists make sense of complex and sprawling negotiations, but has some signature tactics up its sleeves that negotiators watch out for.

One of those key tactics is the Fossil of the Day. Every day, since the 1999 COP in Bonn, members of CAN vote for countries judged to have done their "best" to block progress in the negotiations in the last days of talks. It comes with an elaborate and humorous ceremony chaired by a dinosaur, and which is well attended by media and movement folk.[13] Despite the ridiculous nature of all this, the Fossil has a serious impact on the proceedings of the negotiations, and often leads countries to change their positions to avoid further shaming. Our Japanese colleagues, for instance, regularly report on how their Fossil nominations percolate into Japanese media and government.

Another COP classic is ECO, the civil society newsletter published daily at UNFCCC events that reflect CAN's

---

13. "Fossil of the Day," Climate Action Network International, accessed April 8, 2025, https://climatenetwork.org/resource_type/fossil-of-the-day/.

perspectives and positions on the evolution of negotiations. Published both online and in print copies that are distributed every morning at the entrance to the venue, it is widely read by negotiators. Despite the fact that it is a very "inside baseball" (i.e., highly technical) publication, and that it does not have as large an impact on broader politics as mainstream media interventions, I am always pleasantly surprised to see that negotiators deeply care about its content and whether they feel it well represents or not the state of play and their country's positions. The vivid metaphors and jocular tone are likely a positive contributor. One example of an article that has achieved posterity was one which described then incoming Environment and Climate Change Canada Minister Guilbeault was personally welcomed to his first COP as minister and asked if, given his long history as a CAN member, he would still be a friend of ECO in his official actions.

These tactics do not make headlines in mainstream media. Yet they can be powerful driving forces behind landing specific negotiation outcomes and countries changing their positions, and ultimately, they form the backbone of efforts led by civil society, which create the collective momentum to strengthen constructive diplomacy, and when things go well, ultimately end with more ambitious and inclusive text.

***

Despite the things that are working inside the UNFCCC, and the slow but important progress that is achieved inside the COP space, some things need to change.

Now that the Paris Rulebook has been completed in Baku after Article 6 was adopted, one of the questions that the global climate regime, and we as a climate movement, must grapple with is what it means for parties to move from negotiations to implementation. Despite loads of rhetoric around this, I do not think we collectively have cracked that nut yet.

The proliferation of work programmes and dialogues has not necessarily resulted in the constructive exchange of experiences, ideas, and best practices on the realities of making climate policy work at the national level that these were envisaged for. In fact, the people in the rooms—the negotiators—are competent but often not the right government representatives to contribute expertise on technical implementation questions and influence decisions. Meanwhile there are increasing cost and capacity pressures on an overburdened and increasingly under-resourced UNFCCC Secretariat (which recent defunding from the United States will only accelerate). What is more, delegations—especially those from smaller Global South countries—have only a handful of dedicated but tired negotiators to cover a plethora of agenda items. And perhaps most worryingly when it comes to the actual impact of these spaces, none of the dialogues under the agenda items that are formatted as work programmes have managed to make the transition from dialogue summaries to negotiated text.

With the tenth anniversary of the Paris Agreement in 2025 and the second Trump presidency adopting a "flooding the zone" approach to everything, including international climate policy, there is a convergence of increasing challenges to climate multilateralism at the same time as these reflections on what is working and what is not are bubbling to the surface. Many ideas for reform are floating around, including, notably, from the Club of Rome[14] and the Brazilian COP30 presidency.[15] Some of these proposals aim at better lifting up synergies between the three conventions under the COP,

14. "An Open Letter to the UN Secretary General and COP Executive Secretary: "Reform of the COP process – A Manifesto for Moving from Negotiations to Delivery," Club of Rome, accessed April 8, 2025, https://www.clubofrome.org/cop-reform/.

15. André Corrêa do Lago, "First Letter from the President of COP30, Ambassador André Corrêa do Lago,'" COP 30 Brazil Amazônia, last modified March 15, 2025, https://cop30.br/en/brazilian-presidency/letter-from-the-brazilian-presidency.

focusing on sectoral delivery, and creating other bodies both within the COP and UN systems to close some of the coordination and political driver gaps in the current architecture.

These current times show that, as well as the challenges brought forth by this "polycrisis" also comes opportunity. And, to riff on Gramsci, the old world and its unjust, extractive, and colonial systems are dying. This crumbling of old orders and empires is certainly enabling the emergence of a cabal of hideous monsters. But if we look closely enough, we are also already seeing with it the shoots of a new and hopeful world struggling to be born.

***

COPs capture the complexity of our world, including its destructive dysfunctions. With geopolitical tensions and horrifying genocides ongoing, at a time of rising fascism and crisis convergence, global climate action is more at risk than ever. With five years left before the crucial 2030 milestone, the risk of ceding the COP space to corporate lobbyists and high-polluting industries has never been greater. But the greatest risk would be giving up on global cooperation. That is why I, and Climate Action Network Canada, will continue to attend COPs, to think and rethink our strategies and strengthen our relationships with each other and our international comrades, with governments and their delegations, as we fight for the possibility of a safer, climate-resilient, and saner world.

CHAPTER 12

Municipalities on the Front Line: Local Voices in Global Climate Policy

Berry Vrbanovic and Lauren Touchant

M unicipalities across Canada stand at the forefront of the fight against climate change. This global crisis is felt most acutely in local communities worldwide, and Canada is no exception. For mayors, leadership means being on the frontlines to deal with impact, where the realities of climate change intersect with daily life.

Canada is home to numerous examples of exemplary municipal climate leadership. Over the years, hundreds of municipalities representing more than 80% of the country's population have committed to bold strategies to mitigate and adapt to climate change. Long before provincial and federal governments formalized climate action plans, Canadian municipalities were pioneering innovative approaches. Generations of climate plans emerged at the municipal level, and many mayors have steadfastly championed greenhouse gas reductions since as early as 1992.

This local leadership has reverberated beyond borders. Canadian mayors have forged partnerships and shared their expertise on national and international stages, collaborating with counterparts worldwide through influential climate networks such as C40 Cities, the Global Covenant of Mayors for Climate & Energy, Partners for Climate Protection, and Cities for Nature. Canadian municipalities have played a vital role in shaping the global climate agenda, demonstrating that local action is a cornerstone of meaningful progress.

Located in the heart of southwestern Ontario, Kitchener—the 10th most populous city in the province and the largest within the Regional Municipality of Waterloo—is a medium-sized city, with a population over 300,000, bridging urban and rural diversity. Kitchener's commitment to sustainability has deep roots, stretching back to the 1970s when environmental consciousness began taking shape through grassroots recycling initiatives, long before municipal programs took hold.

This forward-thinking spirit led to a groundbreaking moment in the 1980s when Kitchener became the first community to introduce the blue box recycling program, a concept that would eventually revolutionize waste management across North America. The program's origins, by local resident Nyle Ludolph, were marked by ingenuity and community participation.

The earliest recycling programs in Kitchener were entirely community driven. The Boy Scouts spearheaded an innovative paper recycling initiative, while the Mennonite Central Committee introduced a drop-off program that enabled residents to recycle steel cans, glass, and newspapers. These efforts not only laid the groundwork for broader municipal programs but also showcased the power of community organizations in fostering a culture of sustainability.

By 1983, 35,000 homes were actively participating, with a staggering 75% adoption rate within the first few months of the pilot project. The city set in motion a program that would inspire sustainable practices across North America.

Environmental consciousness is not just a value, it is part of the community's DNA. This deep commitment to sustainability has guided elected officials to take bold steps, establishing Kitchener as a leader in climate action and environmental stewardship.

Back in 1992, Kitchener made history by becoming one of only six Canadian cities to join the newly established ICLEI-Canada office. This was not just a symbolic decision; it was a commitment to action. At a time when conversations around climate change were just beginning to gain traction, the City of Kitchener committed to taking steps to create a more sustainable future.

Through this initiative, the City of Kitchener set out to achieve three critical objectives: Develop comprehensive inventories of local greenhouse gas emissions, produce forecasts for energy reductions and emissions trends, as well as develop tools that would allow municipalities to inform the creation of first-generation action plans to reduce greenhouse gas emissions at the community level.

It is at that time that Mayor Vrbanovic's journey into international climate policy began, when he was first elected as city councillor in 1994. During his tenure at city council, he was deeply involved in the Federation of Canadian Municipalities (FCM), particularly on environmental issues, which eventually led him to engage in international climate work.

Mayor Vrbanovic served for six years on the FCM Standing Committee on Environmental Issues and Sustainable Development, where he contributed to shaping environmental policies and programs nationally. Within this committee, municipalities collaborated on initiatives related to the green economy, climate change adaptation, clean air, and energy, setting a path for more sustainable communities both locally and globally.

In 2006, then Councillor Vrbanovic began to be engaged internationally on behalf of FCM, with his involvement with United Cities and Local Governments (UCLG), leading eventually to his current role as one of the organization's co-presidents.[1] A few years later, he attended COP15

---

1. The UCLG is a global network of cities and municipal association that empowers cities to participate in international governance. It ampli-

in Copenhagen (2009) as one of the FCM leaders. His role was to represent the voice of Canadian cities within global climate governance. As he describes, "Increasing the visibility of municipalities within global climate governance was essential, as local governments have historically been overlooked despite being the first to experience the impacts of climate change."

However, Copenhagen did not offer such a platform: "Copenhagen ended with a big thud; it was not fun as a Canadian."

During COP15 in Copenhagen, Canada was notably awarded four "Fossil of the Day" awards. These are awarded to countries that are perceived as doing the most to delay or disrupt progress in global climate negotiations. Canada picked up a third-place Fossil on the conference's opening day as well as two awards on December 11, 2015, including a first-place award accepted by Toronto Mayor David Miller, who used the platform to call on Prime Minister Harper's government to support a "fair, ambitious, and binding deal," and a second-place award.

Adding to this notoriety, Canada received the infamous "Colossal Fossil of the Year," a title reserved for the country deemed to have the weakest stance in climate negotiations.

---

fies their voices on critical global issues, including climate change. Through the UCLG, member cities can propose policies and frameworks that reflect the interests, roles, and perspectives of local and regional governments in shaping global development agendas, such as the international climate regime. Networks like the UCLG underscore the historical and evolving role of municipalities in international climate governance. The involvement of cities in global environmental policymaking is not new. For example, during the Rio Earth Summit in 1992, cities contributed to the development of Local Agenda 21 (LA21). Incorporated into Agenda 21 as Chapter 28 under Section III, LA21 introduced a voluntary process through which local communities could create and consult policies to achieve sustainable development. This initiative marked a pivotal moment in recognizing the significance of municipal leadership in addressing climate change.

This was attributed to Canada's refusal to strengthen its emissions targets, which were widely regarded as the worst among industrialized nations, and its overall lacking position during the talks.

Reflecting on this reality and realizing the importance that municipalities play in nation building, including achieving ambitious climate goals, then-City Councillor Vrbanovic volunteered to serve as president of the Federation of Canadian Municipalities, from 2011 to 2012. Several years later, he also became a member of FCM's Green Municipal Fund (GMF) Council. The mayor describes GMF as a vital funding program that supports municipalities in their efforts to enhance resilience, enable energy transition toward becoming net-zero communities, and achieve global climate targets.

This dedication to creating sustainable communities became one of the major components of Mayor Vrbanovic's platform as a mayoral candidate. Upon his election, he demonstrated his global commitment to climate action. The municipality of Kitchener joined the Global Covenant of Mayors for Climate and Energy in 2015.

That same year, Mayor Vrbanovic played a pivotal role in global policy advocacy through ICLEI and UCLG during the negotiation of the United Nations Sustainable Development Goals (SDGs). As a representative, his goal was to secure the inclusion of a goal dedicated entirely to cities. Together with local government leaders from around the world, he sought to empower cities in addressing climate change at a time when international agreements often overlooked the critical role of local communities and municipalities.

The adoption of Sustainable Development Goal (SDG) 11 to make cities inclusive, safe, resilient, and sustainable was a significant victory for local government leaders like Mayor Vrbanovic, who had consistently worked through the UCLG and ICLEI to ensure cities were recognized in the international agenda. According to Vrbanovic, for municipalities,

SDG 11 is transformative, directly addressing the unique challenges and opportunities urban areas face:

> Kitchener has fully embraced SDG 11, integrating it into the city's strategic plan. A hallmark of this commitment is the creation of the SDG Idea Factory, inaugurated in 2023. This first-of-its-kind incubator focuses on social, environmental, and equity-based solutions, empowering small entrepreneurs to address challenges affecting both the local community and the planet. Through initiatives like these, Kitchener continues to lead by example, demonstrating how municipalities can drive meaningful progress toward global sustainability.

Mayor Vrbanovic's commitment to climate change and sustainability extended well beyond the SDGs in 2015. As a representative in the FCM delegation, he travelled to Paris alongside fellow Canadian mayors (Gregor Robertson of Vancouver, John Tory of Toronto, and Denis Coderre of Montreal) to meet with Anne Hidalgo, the mayor of Paris, and hundreds of other local government leaders in the midst of the 21st Conference of the Parties (COP) set to start in the city, "Paris was a critical moment for climate. Mayor Hidalgo really wanted this to be a transformational moment. She wanted mayors to be visible and to be heard. She wanted to leverage the 1,000+ mayors' voices and the hundreds of millions of people they represented to influence the goals and the outcomes of the conference."

The program was very structured, with networking events and activities through the Global Covenant of Mayors, including inviting high-level speakers, such as Al Gore. Different strategies and techniques were used including one-on-one meetings, collective meetings, and the organization of a taskforce to help municipalities draft and deliver an effective message.

> While the program is structured, no two days are alike. These conferences provide us the opportunity to use our

voice as a municipality, but we also get to use the power of the collective as a group of municipalities, the voices of those we represent. At all times, we ensure that the voice of communities is front and centre. It is particularly important to do it at the international level because national governments do not always recognize the importance of cities and communities. Opportunities like COP enable us to connect with and encourage national governments that sometimes lose faith in the midst of international negotiations. It is tough work [...].

The purpose of this international gathering of mayors was to rally strong climate support that would include the voices of cities and rural communities. Their advocacy was particularly focused during the final week of the conference, where they worked to persuade reluctant state leaders to sign onto an ambitious Paris Agreement. As Vrbanovic notes, "Each and every day underscores the crucial role that local governments play in the global fight against climate change."

With similar goals in mind, the mayor, acting as one of the co-president of UCLG, also travelled to the UN's COP28 climate conference in Dubai with Kitchener Ward 1 Councillor Scott Davey.

I believe my work with UCLG is important, both for Kitchener and for Canada. We represent the Canadian brand, despite any real/perceived challenges at home. It is important to me because I was 2 and half years old when I moved to Canada. It was the best country in the world, and it remains the best country in the world. We have the ability to make a difference, and we have made a difference in the world through peacekeeping and our work to ban landmines. Cities and communities have a role to play in the image and the reputation of Canada in the world. It does take some investment, long-term investments not within the electoral cycle. This is my commitment to sustainability.

While attending COP28, the mayor also participated in the Climate Action Summit hosted by Mike Bloomberg. "Thanks to mayor Bloomberg, for the first time ever, mayors had a formal meeting in the Blue Zone, the official area where the negotiation happens. The Climate Action Summit program added municipalities to the climate negotiation agenda, particularly on the notion of capacity building."

Mayor Vrbanovic credits former New York City Mayor Michael Bloomberg for choosing to use his personal wealth for philanthropic efforts dedicated to addressing climate change. The City of Kitchener received a US$50,000 grant from Bloomberg Philanthropies' new Youth Climate Action Fund to support youth-led climate initiatives through microgrants in the Canadian municipality, and subsequently received a second grant in 2025 of US$100,000 to support youth grants in our community.

Prior to that, the mayor of Kitchener, together with two city staff, participated in the Bloomberg-Harvard City Leadership Initiative year-long program in 2021–22, further reinforcing his commitment to leadership on city building, including climate action. Kitchener and Mayor Vrbanovic's participation in this program has now been one of 8 cohorts of 40 mayors with a total of 320 graduates, with a ninth cohort just starting this past July. Through this experience, and subsequent professional development opportunities, Kitchener has not only enhanced its own climate policies but also contributed to a broader network of cities committed to innovative and impactful solutions for stronger and better managed and led cities, and their commitment to a sustainable future.

These investments are particularly significant to the mayor, who acknowledges the challenges of the current economic climate. In times of economic hardship, the willingness to invest in climate action often diminishes, even though environmental priorities continue to receive public support. As the mayor of Kitchener reflects,

As a city, through our community engagement and polling, we have learned our residents have always valued the prioritizing of environment and climate change, but, while economic hardship remains high, their desire for investment in this area drops. This creates a challenge for the important work we are trying to accomplish. However, we must never forget that effective change begins at the grassroots level. On the global stage, when states veer off course or fail to recognize the challenges at hand, it is cities and communities at the grassroots level that continue to carry the work forward, keeping these priorities alive. Under Prime Minister Harper, local governments continued their work, much like in the United States under President Trump, in cities like Los Angeles, New York, and others. During those years, local governments kept engaging their communities with support from the international framework. This was especially crucial when national platforms were lacking, though it is less of an issue today under the current Liberal government.

Mayor Vrbanovic's commitment to climate action is firmly grounded in an ethical responsibility, driven by a sense of duty not only to the present generation but also to future ones. "We cannot forget that we have one planet. We are all part of this planet. Engaging communities is critical, so it is about cities and communities across the world. If we don't progress collectively, we are not going to progress at all."

His reflection emphasizes the interconnectedness of humanity and the environment, highlighting the urgent need for collective action. Importantly, he also underscores the economic opportunities that come from building sustainable communities, demonstrating that climate action is not just an environmental imperative but also a catalyst for economic growth. "Beyond the right thing to do, there are huge economic opportunities and employment in cities."

At the core of his argument is the unescapable role municipalities play in enhancing the economic quality of life for their residents while bridging divides. "Cities are particularly important in the time of polarization with many geopolitical issues. These issues highlight the important role cities play in connecting people and communities for peace and sustainability."

This circles back to COPs. International climate conferences are an opportunity to share community stories, engage and connect with federal leaders, and meet with private companies. "Minister Champagne was in Dubai to explore economic development opportunities—as were we. These conferences offer a Rolodex of contacts including in the private sector, which enable us to develop economic opportunities." He continues:

> Networking meetings like COP are very important. Let me take the example of a meeting organized by the former Ontario Premier Wynne in Toronto. Former Vice President Al-Gore was there. During the event, I was able to meet with the president of IKEA, he talked about sustainable development, we had a subsequent meeting, and IKEA now has a store in our community. Local governments are uniquely positioned to bring people together beyond political cleavages. They are also uniquely positioned to drive change, foster innovation, create green jobs, and improve the resilience of communities to the impacts of climate change. By embracing sustainability, municipalities can unlock new economic potential while ensuring a high quality of life for all residents. This approach offers a clear vision: sustainable development is not just about environmental stewardship, but about economic vitality and a thriving, resilient community.

Mayor Vrbanovic's perspective reinforces the idea that tackling climate change can—and should—go hand in hand with building a prosperous, peaceful, sustainable future for all, as outlined in SDG Goal 11. Therefore, the future of COP must involve changes, especially in recognizing local governments as a permanent and integral part of the process. For Vrbanovic:

> We need a permanent seat at the table, but some national governments don't want this to happen. Yet, we are the order of government closest to citizens, so we should play a more important role at the national but also international levels. We are the first one impacted, but we are also the first ones to implement the solutions. We were not always welcome at the United Nations (UN), the UCLG could not get a room within the facility, we had to get a hotel room to conduct our events, now the UN does see the value of the municipal governments. Current Secretary-General Antonio Guterres, increased opportunities for cities and communities to have a voice at the table. We were successful in 2015 to get a SDG goal. Everyone recognized the role of cities. 60% of global emissions are emitted by our cities. A similar recognition must occur in the context of the UNFCCC.
>
> [In Canada], there's been a change with the current government, we were able to be part of the official delegation. The government has used cities as a tool and as partners to tell the story. One challenge remains, some Premiers have been vocal about the participation of cities in the delegation. We deal with a 19th-century framework that does not work in the 21st century. I don't think Canadians buy that. They see three orders of governments, they don't care who does what, or who pays, as long as it improves their lives and the lives of their families. It has been more effective since the federal government has been working directly with municipalities through the FCM. It has resulted in better outcomes. I get the political realities, but the bickering and the in-fighting needs to end. We all have limited resources; we need to be barn raising to achieve greater outcomes for all Canadians.

Canadian municipalities and cities around the world have an important role to play. They will continue to get involved in future COPs regardless of political changes. "We must say united and speaking with one voice, has been the strength of municipalities."

But he has one last message for all of us:

Canada is an amazing country that has a strong history, without the insights of successful decisions or decisions that could have been better. We are a country coming together, we take the best of each other, we thrive on multiculturalism. I am a kid from a dictatorship that has had the opportunity to become a mayor, to serve my neighbours. We have the opportunity to shape a response, to shape the future of the planet and ensure that our vision is not shaped by others. I invite all Canadians to embrace this vision, to participate and be engaged at the community level, this is where we are going to be building the future.

CHAPTER 13

Cops Are Slow,
But Cities Are on the Go

David Miller and Lauren Touchant

I have attended three Conferences of the Parties (COPs), each in a different capacity. My overwhelming impression is of the contradictory nature of these events—the incredible importance of the world coming together annually to set out a path to address an urgent existential challenge, together with the slow progress that actually has happened; the failure to reach full agreement in 30 years of negotiation together with the Paris Agreement's framework for responsibility; and most of all, the juxtaposition of a process that in many ways represents failure with the excitement and optimism of those attending each year, particularly civil society.

COP15, Copenhagen, 2009
Lessons learned:
- The requirement for consensus makes international climate action exceptionally difficult.
- The role of civil society and the youth movement is essential.
- Cities lead climate action.

I attended COP15 in Copenhagen as the mayor of Toronto and the elected chair of C40 Cities, then an organization of mayors of 40 of the world's most significant cities like

London, Paris, New York, Cape Town, Beijing, Buenos Aires, and, of course, Toronto. COP15 was intended to be the COP where agreement was reached, and was promoted as such. It appeared to have the right ingredients—a host in Denmark who was both a climate leader and an influential country diplomatically; timing given a supportive US president; and necessity. But some were prescient that the COP might not succeed—for example, C40 and the mayor of Copenhagen co-hosted a climate summit for mayors whose slogan was "While Nations Talk, Cities Act."

The COP was held at a convention centre outside Copenhagen, near the airport. Its atmosphere was sterile and serious—except for the significant presence of youth delegates and civil society, who were loud, raucous, and determined. Canadian organizations, for example, had arranged awards, serious and satirical, media coverage of key issues, and brought musicians such as Sarah Harmer to COP to both raise the spirits of the attendees and make a point themselves. Their advocacy was a vivid contrast to the evidently highly organized but discordant proceedings inside the COP itself, where attempts driven by the United States to reach agreement—the so-called Copenhagen Accord—backfired, even though it contained a breakthrough in reaching agreement with China. The blame generally has been given to the fact that the United States brokered an agreement that excluded a broad range of countries—a fair criticism—but also obscured the truth that a number of countries were using the principle of consensus that governs the COP process to delay action, as, for example, Canada.

The former prime minister, Stephen Harper is a right-wing climate sceptic who was reliant on Canada's oil patch for his electoral base and funding. While a Conservative, he comes from the Reform Party—a western populist party—that merged with Canada's Progressive Conservatives, a moderate centre-right party, and essentially took over the party completely, moving it significantly to the right. By COP15, Harper was confident in his anti-climate agenda but sneaky about how he accomplished it because he knew that Canadians were broadly supportive of climate action. His first steps were to withdraw Canada from the Kyoto Protocol and then adopt new climate targets that sounded adequate to an uninformed observer—but actually embedded an increase in emissions.

At COP15, Canada received significant criticism for delaying the process—delays the host Denmark seemed unable (or unwilling) to deal with. Because international negotiations like this operate on consensus, and where there are disputes, a common practice is to bracket draft language that is causing difficulties and return to it later. Canada gained a reputation at this COP—and in the run-up to it—for bracketing unnecessarily over matters that were not really in dispute, just to cause delays. I was told that at one point they had bracketed a comma. True or not, it reflects the tactics of the Canadian government, and because the negotiations are not public, it allowed the government to pretend that it was only raising reasonable concerns for an oil-exporting nation, when in fact it was doing everything it could to undermine the negotiations as a whole.

Civil society noticed, awarding Canada several "Fossil of the Day" awards. As one of the most senior Canadian elected officials present in Copenhagen, I was asked to receive the award and did so, saying I was embarrassed as a Canadian. Don Wanagas, my excellent director of communications, had briefed me before the award and used that word to describe his feelings. He did not tell me to use it—Don did not advise me that way, preferring to brief me and let my own instincts take over—but it summed up how I was feeling too. *The Globe and Mail*, a leading conservative newspaper, then ran an editorial criticizing me for both using those words and for attending the COP. It was an outrageous editorial, as the appropriate criticism, given the seriousness of climate change, should have been aimed at the Canadian government and Prime Minister Harper. I wanted to respond, and Don contacted them to ask for a right of response. They offered a live one-hour, midday, online text conversation with Globe subscribers, which we accepted. During the session, without advising us that they were going to do so, they ran a poll phrased approximately as "Mayor David Miller says he is embarrassed by the Canadian government's actions at COP15. Do you agree?" I was concerned that the paper had done this without our agreement, but Don told me not to worry, that *Globe* readers (who skew conservative) would agree with me. And it turned out they did—something like 70% shared my view, making the online session a political triumph. When I asked Don why he had been so certain, he revealed that he had noticed a similar poll a few days earlier in the *Globe* with similar results, which was why he had mentioned the word "embarrassed" in the first place. He anticipated the *Globe's* reaction, that I would be able to respond, and, further, the ultimate response of its readers.

The contrast between the formal negotiations and the role of cities could not have been clearer. There was heavy security at the negotiations, they moved at a slow pace, and news was glum. The Copenhagen Climate Summit for Mayors, in contrast, was open to the public, exciting, and all about climate action. My co-host, Copenhagen Mayor Ritt Bjerregaard, was the former European environment commissioner and a smart and savvy leader. Between us, over 100 mayors were invited to the square in front of City Hall to present their cities' climate action. The energy was incredible—mayors like New York's Michael Bloomberg and Mexico City's Marcelo Ebrard made strong commitments in line with what science required and the Kyoto Protocol, and cities like Cape Town and Melbourne spoke to action already underway. But in addition to the bold action, there was a strong sense of civic engagement—and hope. Mayor Bjerregaard had commissioned from MIT a new type of pedal-generated electric-assisted bicycle for the occasion— to this day, searching online for "The Copenhagen Wheel" can lead you to a video of multiple mayors riding through Copenhagen joyously with at least one singing the Beatles song "Eight Days a Week," albeit slightly off-key. Members of the public were welcome and encouraged to come to the events at City Hall—and it was easy and simple to get in. The Copenhagen Climate Summit for Mayors was so successful and created so much buzz that city-led climate action became the front-page story on the daily newspaper created for delegates to the COP itself—making it clear where the action was, and sadly, where it remained given other events at the COP.

COP21, Paris, 2015

Lessons learned:
- Leadership matters.
- Details matter.
- Cities matter.

My role at COP21 was significantly different from Copenhagen, as I attended as the head of a Canadian non-profit—the World Wildlife Fund (WWF Canada, the Canadian affiliate of the World Wide Fund for Nature). In that sense I was more of an observer than a participant, but, of course, it was an incredible COP in which to have the privilege of observing.

The first thing that was clear was the attention to detail by the host country, France. It had laid the groundwork well over the previous year and there was good momentum leading up to the COP. No doubt there were many measures taken by the French government to keep that momentum going, but I will highlight three that I was aware of and that seemed the most important to me:

1. The French President took an active interest in the negotiations and lent the weight of his office to assist, not just at the COP but in the year long run up to it.
2. French negotiators assumed control of the process and did not leave the text to chance. As the deadline for agreement approached, each evening the French rewrote the text to accommodate all reasonable positions, shortening the document and thereby forcing a transparent and honest

conversation about real areas of disagreement. In doing so, they circumvented tactics like those used by Canada in Copenhagen six years earlier and forced the process to work.

3. They realized that it was important to treat the negotiators with respect, and ensure, as far as possible, that they were happy despite long hours and tense negotiations. One way they did this, which led to numerous comments about the French stereotype, was to ensure excellent meals. Delegates raved about the food, and in the civil society zone—the Blue Zone—we benefitted too. The organizers arranged for small coffee trucks with actual espresso machines to be driven onto the floor. Not only was the food good, but the coffee was exceptional, as good as any I have ever had, anywhere—not normally what one expects in a convention centre. It is easy to joke about the French and their love of food, but ensuring delegates had everything they needed—including superb meals—was a subtle but smart and thoughtful step.

Cities were present too, in massive numbers. Paris Mayor Anne Hidalgo, now chair of C40 Cities, with the support of Bloomberg Philanthropies and Mike Bloomberg, as special climate envoy to the UN secretary-general, organized the Climate Summit for Local Leaders at Paris City Hall. Huge numbers of mayors attended—hundreds, with notable interventions by mayors of cities as diverse as New Orleans, Amman, and Seoul. The event gave a sense of momentum to climate action, followed as it was by numerous events in the Blue Zone at the COP itself, featuring city actions, with mayors, city councillors and their advisors, and C40 Cities staff as speakers.

As an NGO, WWF Canada always pays close attention to science—we saw it as our duty to be truthful and to speak truth to power. So when the Paris Agreement was signed, we had a dilemma. By 2015, there was little doubt that the goal for the world needed to be to hold overall average temperature rise to 1.5°C—but the Paris Agreement goal was 2°C (with a "higher ambition of 1.5" negotiated by Canada's then minister for the environment, Catherine McKenna). In addition, because the mechanism of the Paris Agreement was to allow each country to define its own contributions, with a built-in system for ratcheting up these commitments at future COPs, if collectively insufficient, it was not exactly the "legally binding" agreement that activists had been demanding for years. What is the duty of a science-based organization in these circumstances?

Like nearly all others, we decided to overlook these flaws and focus on the extraordinary fact of the Agreement itself. It had taken 21 years, and without the incredible leadership of the French, likely would never have happened. And the fact of an agreement led to potential momentum, for example in the business world, as climate action became normalized globally. And we all hoped that there would be future acceptance of 1.5°C and increased action at subsequent COPs. We were right on the momentum, and on 1.5°C—for which C40 deserves significant credit—but, as events have shown, less so on the possibility of success at subsequent COPs. "We will always have Paris" is not just a great line from a film—it is the one firm foundation for multilateral climate action.

COP26, Glasgow, 2021
Lessons learned:
- International diplomacy is structured to make incremental gains—moments like Paris are rare.
- City leadership can influence significant climate action—and ambition.

By the time of the Glasgow COP in 2021, I was in a new role. I had returned to C40 Cities, now significantly larger, both in number of cities (nearly 100) and staff (nearly 200), and was the managing director of diplomacy. It was my role, and that of our excellent team, to support the collective efforts of C40 mayors to influence international climate action. In this sense, Glasgow was a triumph. If in Copenhagen the role of cities as meaningful actors on climate change was established, Glasgow conclusively demonstrated their success in also influencing multilateral processes.

Led by C40 Chair and Los Angeles Mayor Eric Garcetti, supported by Chair-Elect London Mayor Sadiq Kahn, and hosted by the Leader of Glasgow Council (equivalent to a mayor but in a parliamentary system) Councillor Susan Aitken, mayors, city advisors, and C40 staff were everywhere, speaking to technical and political issues. C40 mayors arrived in Glasgow by train from London—giving journalists a unique opportunity to interview them ahead of the COP, and in many ways setting a news agenda for the following two weeks. Glasgow, under the able leadership of Councillor Aitken, hosted a two-day summit at city hall, showcasing the latest action and ambition of cities. More importantly, mayors and other city representatives were invited to numerous business events—from financing electric buses to clean construction to architects and engineers' association forums on green buildings, an explicit recognition of the importance of city-based policy and action to greening the economy. Mayors met with the UN Secretary General to discuss his ambitions and how cities can help—and also how the international system could assist cities in turning their ambition into faster action. They also met with venture capital, banks, philanthropic donors, civil society, youth delegates, the COP president, and others. Mayor Garcetti, with an alliance of trade unions, businesses, and others, launched the Global Green New Deal campaign—publicly connecting the need for jobs, to address inequality, and a just transition with climate mitigation. Significantly, the cities' race to zero was launched by Mayor Garcetti on the floor of COP—the first time a non-national politician had been invited to make an official announcement there. The cities' race to zero, launched by the Climate Champions in partnership with a group of city-based organizations convened by my team at C40, had recruited, at Mayor Garcetti's instigation, over 1,000 cities to commit to the same high standards as C40 cities—and to the actions necessary to undertake them.

And it is those standards that in some ways represent the most remarkable achievement of cities—changing the ambition of the COP process itself. As of 2015, C40 had nearly 100 cities with a population of over 600 million in their urban region—today, it is close to 700. The collective impact of their actions is akin to a very large country. So when the cities take collective action, it has a significant impact globally—and, if handled correctly, diplomatically.

In 2016, Mayor Eduardo Paes of Rio, C40 chair, tasked C40 staff with developing a report showing what C40 Cities needed to do to implement the Paris Agreement—at 1.5°C. At that year's C40 Summit in Mexico City, C40's elected Steering Committee of Mayors unanimously endorsed the report—known as Deadline 2020—which for the first time developed trajectories for cities to do their fair share of reducing carbon on a 1.5°C path. Roughly, cities in the Global North needed to peak emissions by 2020 and do their fair share of halving by 2030 on a path to net zero by 2050 (the timeline for peaking for Global South cities was slower). By Glasgow, a huge percentage of C40 cities had made this commitment, developed the necessary plans, and begun implementing them. Toronto, for example, had its first climate plan in 2007 and by 2022 was 33% below 1990 levels in greenhouse gas emissions for all activities, private and public, in the city.

In our diplomatic activity between 2016 and 2021, C40 added our voice to the growing calls from scientists and civil society for a 1.5°C target to be reached. We were able to add a layer to the advocacy that was enormously effective—by showing that governments representing hundreds of millions of people could not only agree to such targets but could realistically implement those plans with meaningful action—on transportation, buildings, electricity generation, waste management, urban development, food, and much more.

As is well known, by the time of Glasgow, the 1.5°C goal became an accepted standard—almost without debate. The UN reports the outcomes of that COP this way:

> Nations adopted the Glasgow Climate Pact, aiming to turn the 2020s into a decade of climate action and support. The package of decisions consists of a range of agreed items, including strengthened efforts to build resilience to climate change, to curb greenhouse gas emissions and to provide the necessary finance for both. Nations reaffirmed their duty to fulfill the pledge of providing 100 billion dollars annually from developed to developing countries. And they collectively agreed to work to reduce the gap between existing emission reduction plans and what is required to reduce emissions, so that the rise in the global average temperature can be limited to 1.5 degrees. For the first time, nations are called upon to phase down unabated coal power and inefficient subsidies for fossil fuels. As part of the package of decisions, nations also completed the Paris Agreement's rulebook as it relates to market mechanisms and non-market approaches and the transparent reporting of climate actions and support provided or received, including for loss and damage.[1]

For this we must primarily acknowledge the scientific research and the IPCC, but the contribution of the world's largest cities in adopting such standards and in showing that they work—technically, politically and economically—cannot be underestimated. It mattered.

---

1. "The Glasgow Climate Pact – Key Outcomes from COP26," United Nations Climate Change, accessed July 23, 2025, https://unfccc.int/process-and-meetings/the-paris-agreement/the-glasgow-climate-pact-key-outcomes-from-cop26.

CHAPTER 14

From 2015 to COVID, We Were Still Dreaming

Patrick Rondeau

As far as I can remember in my trade union career, climate issues have been for me a driving force for change. After 25 years of unionism, I realize how much the COPs have changed my trajectory and that of the Fédération des travailleurs et des travailleuses du Québec (FTQ). Already in the early 2000s, as a simple activist, I was submitting resolutions at FTQ conventions to adopt political proposals regarding climate change. My main motivation was my daughter, born in 1999.

Without knowing the work of Tony Mazzocchi, an American trade unionist, on just transition, it seemed appropriate and wise to address climate and labour issues in a complementary way rather than in opposition to each other.[1] However, at the time, these issues were black or white. Of course, over time, I realized there were shades of gray, but I

---

1. Annabel Pinker, *Just Transitions: A Comparative Perspective* (The James Hutton Institute & SEFARI Gateway, 2020), 8, https://www.gov.scot/binaries/content/documents/govscot/publications/independent-report/2020/08/transitions-comparative-perspective2/documents/transitions-comparative-perspective/transitions-comparative-perspective/govscot%3Adocument/transitions-comparative-perspective.pdf.

was unaware of them. When it came to blocking shale gas, the Gaz Métro union would stand up and say, "Wait a minute, jobs first!" When it came to discussing nuclear power, the union at the Gentilly plant would stand up and say, "Wait a minute, jobs first!" And so on. Yet, the first climate change demands at the FTQ date back to the 1960s.[2] But until the mid-2010s, unions simply did not feel the effects of climate change or its emergency.

Nevertheless, the FTQ Montreal Metropolitan Regional Council (CRFTQMM), where I was a member of the executive board and then regional advisor for the FTQ, started thinking and, gradually, it was no longer a question of accepting the environment/labour opposition. Work was initiated through reflection days on eco-socialism and then on the fight against climate change. It was clear that the climate literacy deficit within the union ranks was immense. Gradually, the CRFTQMM began to question decisions at FTQ general councils, including support for the reversal of Enbridge's Line 9. The CRFTQMM wanted to provoke a discussion on this resolution to elevate the debate within the federation. The resolution was still adopted but mandated with new conditions, including the application of the highest safety standards. This was in 2013.

In the meantime, my desire to elevate the debate intensified, better understanding union structures and political issues. Added to this were alter-globalization issues. Climate is linked to economic and social conditions. Development plans differ from one country to another, and the approach must be global while having an impact at the local level. Very quickly, it became evident to me that social and environmental struggles are intrinsically linked. It was in 2012

---

2. Colin L'Ériger, "Changeons le Québec, pas le climat - Déclaration de politique sur les changements climatiques," 31st congress of the FTQ, Montréal, Quebec, November 28 to December 2, 2016, 7, https://ftq.qc.ca/wp-content/uploads/2016/12/Declaration-politique-climat-FTQ-Congres-2016.pdf.

that the FTQ began to take a serious interest in the World Social Forum (WSF). As a regional advisor for Montreal at the FTQ, I was approached by Alternatives and the YMCA to join a Quebec delegation to the Rio + 20 Summit in Brazil. We organized a very small FTQ delegation and left for Rio de Janeiro. It was a brutal awakening to global issues, forms of mobilization, and the need to think and organize beyond Quebec's borders. Then, WSF editions followed (Tunisia twice, Brazil, and Montreal). Gradually, the FTQ went from participant to co-organizer of Quebec delegations, which each time included at least 60 people from various civil society organizations and universities.

Then came 2014. Exhausted by Stephen Harper's neo-conservative policies, Canada's progressive organizations gathered to organize a Canadian people's social forum (PSF) in Ottawa. The FTQ was at the centre of the organization, and in the meantime, I had added the climate change file to my workload, an orphan file at the FTQ for several years. At the same time, the same organizations coordinating Quebec delegations to the WSF approached us to form a delegation for COP21.

The PSF was a huge success in the summer of 2014, and the FTQ quickly advanced on the climate file. First, by pacifying its relations with Greenpeace. In 2013, the environmental NGO and the Grand Council of the Crees filed complaints against the forestry company Produits forestiers Résolu for "forest crimes" with the Forest Stewardship Council (FSC). These actions resulted in the enterprise losing its FSC certification. These actions had provoked the ire of actors in the Saguenay–Lac-St-Jean region, where the enterprise was based, openly opposing workers and environmental groups, mainly from Montréal. The relationship between Greenpeace and the FTQ initially took the form of courtesy in informing each other of public outings. Then, a collaboration was established where each organization supported the other, recognizing their respective areas of expertise. It was at this moment that the labour and environment equation began to change, becoming a complementary narrative.

Then came COP21 organized in Paris in 2015.

Where to start?

First, by mentioning that the very principle of COPs was totally foreign to me and the FTQ. We had no idea what it was about. However, the FTQ recognized the aspects of participating in the WSF. First, it allows updating certain practices such as communications or mobilization. Then, the WSF offers a unique international networking space, allowing for a deeper understanding of different realities. So, it was not the attraction of the COP itself that brought the FTQ to Paris, but rather the dynamic of this civil society delegation found in the WSF. We took part in the adventure of co-organizing a Quebec delegation for COP21. Moreover, civil society organizations were organizing a multitude of events on the sidelines of this COP. Paris was buzzing with the energy of COP. Conferences, seminars, spaces, engaged shows, and fairs organized by civil society took place every day during COP21.

Then, a few days before the opening, the Bataclan tragedy occurred.[3] Deeply shaken, we nevertheless took the plane to Paris. This event also changed the trajectory of this COP and the interpersonal and professional relationships of the Quebecers present in Paris. A solemn and very human solidarity was established. Then, the hope of being able to change the world's trajectory. I was also amazed at the magnitude of the climate crisis and the projections of global warming between 4°C and 6°C by the end of the century. What we learn at the COPs transforms us.

The FTQ managed to obtain three accreditations through the International Trade Union Confederation (ITUC). So, we were two advisors accompanying the FTQ's then-general secretary. To be honest, we were lost. We wandered through the

---

3. "Attentats de 2015 en France (dont Charlie Hebdo et 13 novembre)," *Le Devoir*, https://www.ledevoir.com/attentats-france-2015.

long corridors, trying to understand the texts, discovering the Climate Action Network Canada (CAN-Rac), the trade union constituency (TUNGO), and the North–South dynamics seen by states and activists.

Each evening, we returned to the hotel and tried as best we could to explain what had happened to the other members of the Quebec delegation, who, in turn, explained what had happened in the various civil society activities. It was very clear at that moment that one could not go without the other. Mobilization must be internal (lobbying, political meetings, etc.) and external (mobilization, conferences, etc.).

We also held a series of meetings with Quebec's political parties. The Parti Québécois (PQ) was headed by Pierre-Karl Péladeau, and Manon Massé represented Québec solidaire. Quebec's minister of the environment, David Heurtel, cavalierly refused to meet with us. This dimension of exchange with political parties was clearly an avenue worth exploring, but we still needed to know what to ask them in concrete terms. That is when we learned about the concept of just transition, as proposed by the International Labour Organization (ILO).[4]

A real revelation for the FTQ. Not only are we joining the fight to land this concept in the Paris Agreement, which will end up in the preamble, but we quickly understand that this will become our hobbyhorse for years to come. It is no longer a question of defining union action in the fight against climate change, but rather of adapting these principles to the realities of Quebec, that is the regions and economic sectors, according to what has been agreed by unions, employers, and states, under the ILO. These principles are based on four pillars: social dialogue, social protection, decent work, and labour

---

4. International Labour Organization, Guidelines for a Just Transition Towards Environmentally Sustainable Economies and Societies for All (ILO, 2015), https://www.ilo.org/sites/default/files/wcmsp5/groups/public/%40ed_emp/%40emp_ent/documents/publication/wcms_432859.pdf.

rights. Add to this the need for equity, then industrial, social, and public policies, and we have everything we need to begin a colossal task in Quebec, enabling the FTQ to assume a leading role in the fight against climate change.

The Paris Agreement was thus adopted, to great applause, but it soon became clear that the work to be done to implement it and achieve its targets was abysmal! And how could we forget the Trudeau government's "Canada is back!" claim. Freshly elected with its sunny ways, Canada shone on the international stage. It was multiplying its efforts to rally the sometimes-far-flung positions of developed and developing countries and was assuming an undeniable leadership role, which was applauded, not the least by its own civil society. Moved, the latter gave the negotiating team a standing ovation.

It is therefore largely thanks to the results of this COP that the FTQ has deployed its work on just transition. As early as 2016, it organized a tour across Quebec to initiate a dialogue with its members. This tour validated the level of literacy concerning climate change, exposed the dangers of this crisis for the economy, and explained the concept of just transition. These sessions concluded with the drafting of a policy statement, which was tabled at the FTQ convention in November of the same year. This policy statement became the FTQ's North Star in the fight against climate change and just transition.[5]

The FTQ did not take part in COP22, not fully understanding its role in this ecosystem, but witnessed Canada's announcement to shut down its coal industry by 2030, without a precise just transition plan. A social and economic crisis quickly ensued in Alberta the year following this announcement, as the coal industry immediately began shutting down many of its operations, without waiting for a plan to be shut down until 2030. Implementation of a just transition was too late. This landmark event crystallized Canada's relationship with the just transition concept to this day.

---

5.   Colin L'Ériger, "Changeons le Québec, pas le climat," 7.

The coal misadventure was one of the reasons for the FTQ's return to the COPs. It had become clear that the labour movement needed to come back to the fore to better adapt and implement the concept of just transition to avoid worker resistance to the changes that were needed. It was also clear that the concept was perceived solely as a response to job losses. In fact, the just transition concept aims to prevent job losses, improve the living conditions of workers and their communities, involve those affected in the various transition plans, and initiate a serious rethink of our economic system.

It was in Bonn that a small FTQ delegation (two people) travelled to attend this COP under the presidency of the Fiji Islands. It was during this COP that the FTQ was able to explain the just transition concept to Sylvain Gaudreault, then a PQ deputy. Gaudreault went on to become a true spokesperson for the just transition concept, as defined by the ILO, in the House of Commons, until his retirement from politics:

> The FTQ has done a tremendous amount of work on the Just Transition concept. Many of its members are workers in the industrial sectors most affected by the ecological transition. In my opinion, union leaders who choose to tackle the just transition issue head-on in their bodies, when the majority of their members come from the exploitation of raw materials, the petrochemical sector or the world of construction, are showing great courage.

However, it was also during this COP that the trade union movement realized that the just transition was not embodied in the negotiations. Together with the ILO, the trade union movement organized conferences to inform delegates. The concept of climate negotiations became a little clearer for me, but nothing more. The language of the UN Framework Convention on Climate Change (UNFCCC) is arid. What is more, the link between trade unionism and international climate negotiations makes sense but does not attract tons of delegates from trade unions.

It was not until 2018, at COP24, that the concept of just transition began to take off. The Polish presidency decided to open the COP with the Silesian Declaration on Solidarity and Just Transition.[6] Silesia is a region of Poland that has had a hard time getting out of coal. A very interesting case study for those interested in energy transition from a just transition perspective.

But then, as hardly anyone knew the ILO's guiding principles, let alone the organization itself, the dance of interpretations of what a just transition should be began. It is true that the term can be interpreted in many ways. With a delegation that is clearly smaller than the other UNFCCC constituencies, that is one of the nine civil society groupings recognized by the Convention, and with less negotiating experience overall, the trade union movement is trying, at its best, to preserve all the work begun since 2013 and the basic elements that constitute just transition. The work is arduous and difficult and puts the trade union movement in a defensive position that does not allow for collaborative work.

Meanwhile, in the wake of the Paris Agreement, the forum on the implementation of response measures had already integrated the just transition concept into its reflections and was attempting to finalize its work plan. It was during this COP that the Polish presidency created the Katowice Committee of Experts to support the forum's work. It is this work by trade union non-governmental organizations (TUNGOs), the ILO, and the forum, therefore, that kept the just transition guidelines on track. However, it was becoming clear to me that TUNGO could not remain isolated for long.

---

6. Just Transition Solidarna Transformacja, "Solidarity and Just Transition Silesia Declaration," COP24, 2018, https://www.ioe-emp.org/index.php?eID=dumpFile&t=f&f=134978&token=91237abd5b4e38c-1e7c2e4364b2b8e7095d8e0fd.

The year 2019 opened a new door: that of the sessions of the subsidiary bodies, the Subsidiary Sessions (SB) in Bonn, every June. The ITUC was looking for someone to represent it at the 2019 session. My management had authorized me to attend. I was the only one representing TUNGO. I was doing my best, but it was clear that this space also needed to be better invested in, since unlike the COPs, the SB session in June brings together fewer people and the learning curve is therefore easier. What is more, this session is an important milestone before the next COP. The proximity of the participants also enables constructive exchanges between the parties and civil society organizations. In fact, the big difference between COPs and SB sessions is that it is only in COPs that new commitments can be negotiated. At the SB session in June, we make progress on decisions that have already been adopted.

Then, in 2019, a social crisis erupted in Chile. An excessive rise in the cost of public transport tickets enraged the people, who were fed up with social policies that were impoverishing them. They took to the streets *en masse*, even to the point of overthrowing the government. Chile was due to preside over COP25. A concrete example that demonstrated that the climate crisis could not be resolved in the presence of a social crisis, and that the reverse is also true. They are two sides of the same coin. It also goes to show that, although policies come from governments, it is civil society that will have to implement them.

This puts the whole principle of climate negotiations under the UNFCCC into perspective. Although it is a party-led process, civil society and the UNFCCC's official constituencies cannot be ignored in the negotiations. The Convention is equally clear on this point, with observers having a prescribed role[7] of active participation, with the aim of contributing to the climate convention, that is to stabilize global

---

7. UNFCCC, *Observer Handbook for COP29* (UNFCCC, 2024), 22–23, https://unfccc.int/sites/default/files/resource/Observer%20Hand book%20for%20COP29%203008%20pub%20%281%29.pdf.

warming and reduce its effects. The Intergovernmental Panel on Climate Change (IPCC) has also been very clear in the last 10 years, underlining the importance of civil society participation in the processes, if countries wish to give themselves the slightest chance of success.

But where is Canada in all this? It is staying the course, assuming a less flamboyant leadership than in 2015, but just as much. The link between the Canadian government and its civil society has been maintained. The latter continue to ensure that they meet with the various Canadian organizations throughout the COPs. There is real collaboration.

14.1. A Crisis of Confidence

Then came the pandemic.

This health crisis clearly demonstrated the limits of our economic system, but above all, its lack of resilience. It also showed the limits of international cooperation and the extent to which North–South disparities are exacerbated. The situation of vaccines is a case in point. We knew we were not all equal, but we watched helplessly as millions died to save others.

Then came the strange episode of the COP being put on pause, like the rest of the world. The first SB session during this pandemic took place virtually, over three weeks, in June 2021, with a rotating schedule. This event once again demonstrated the inequalities between North and South. Developed countries were able to participate easily in the various negotiation agendas, while developing countries, often losing Internet signal, were clearly at a disadvantage and even penalized.

What is more, it was clear that the "connected" negotiators were annoyed with those who had difficulty expressing their views because of network failures. It was the best demonstration that voices are not equal. It also answered a crucial question: Why not organize the COPs virtually? Quite simply because it does not work, and some people are systematically disadvantaged. Nor can we conclude that the results of this SB session were the best. For the most part, it has led to a stagnation of agendas.

The pandemic also meant a change in my career path in the COPs. COP25 was the last one I attended as an observer. The ITUC asked me to take on the role of co-coordinator of TUNGO by becoming a co-focal point, that is to contribute to the organization, support, and supervision of international trade unions during the work of the UNFCCC. I accepted. This new role has given me a different perspective not only on the negotiations, but also on all the UNFCCC's mechanisms and political power plays, while still being on the side of observers.

The years following the pandemic were very different. First, COVID has left deep scars that are difficult to heal. There was the spreading economic and social crisis. We are moving from a climate crisis to a context of "polycrisis." As mentioned earlier, these crises must be addressed at the same time, otherwise they cannot be resolved. Then, as global warming continued, all indicators were in the red. Planetary boundaries[8] are being exceeded one after the other, and greenhouse gas (GHG) levels in the atmosphere are rising.[9] Since 2015, hope lay in the fact that the needle was moving. In other words, it has gone from 4°C to 6°C, to land between 2.4°C and 2.9°C in the post-pandemic years, and only if the commitments made by governments are respected. But now it is stagnating and even starting to move again, but in the wrong direction. Why is this? The first answer is the inability of fossil fuel-dependent countries to switch to renewable

---

8. "Planetary Boundaries," Stockholm Resilience Center, accessed July 8, 2025, https://www.stockholmresilience.org/research/planetary-boundaries.html#:~:text=The%20Planetary%20Boundaries%20are%20the,nine%20boundaries%20have%20been%20transgressed.

9. Environment and Climate Change Canada, *Global Greenhouse Gas Emissions: Canadian Environmental Sustainability Indicators* (Environment and Climate Change Canada, 2024), 5, https://www.canada.ca/content/dam/eccc/documents/pdf/cesindicators/global-ghg-emissions/2024/global-greenhouse-gas-emissions-en.pdf.

energies for their own energy production. Even a country like Canada cannot escape this and instead tries to use all sorts of stratagems to postpone the inevitable reality. Even today, more than 80% of energy production is fossil-fuel-based. This creates the delicate situation in which the production and sale of energy are intrinsically linked to the global economy. Changing the way we use energy means changing our economy, and vice versa. States are still incapable of doing this.

Something even more insidious is creeping into the dynamics between states: a crisis of confidence linked to climate finance.

The Paris Agreement renews the financial commitment between developed and developing countries. To offset their emissions, the countries of the North must now pay $100 billion a year to the countries of the South, between 2020 and 2025. This agreement will not be fulfilled until 2023, three years later than planned. The issue of climate finance will monopolize discussions in the coming years. First, the very concept of climate finance will be challenged. That is, the obligation of developed countries to financially compensate developing countries for the disastrous consequences of their GHG emissions, and the obligation of the latter to use the sums paid strictly in accordance with the agreements adopted at the COPs. The whole concept of international cooperation is also challenged. How can we ensure that international financial aid does not simply become a way of enriching contributors? In other words, how can we ensure that the sums paid out by developed countries to developing countries do not simply become calls for projects with a return on investment, thereby increasing the debt burden of certain countries in financial difficulty?

Then there are always the frustrating loopholes. For example, the Glasgow Pact introduces language on fossil fuels for the first time, but only soberly.[10] The real focus is

---

10. *Report of the Conference of the Parties serving as the meeting of the Parties to the Paris Agreement on its third session, held in Glasgow from*

on financing. This leads to aberrations such as requiring developing countries to stop funding their fossil fuel exploration and exploitation, and to turn immediately to renewable energies, without which the sums will not be forthcoming. However, these same countries, including Canada, continue to explore and exploit their fossil fuels, since the Glasgow Pact specifies that all inefficient financing must cease, without any clear definition of what this means. In Canada's case, the evaluation of efficient financing is largely done with the Finance Department.[11] This includes tax credits for oil companies that commit to decarbonizing their operations through CO_2 capture and storage. To date, these processes have yet to prove so effective that they would enable Canada to achieve carbon neutrality by 2050. Worse still, Canada is forecasting an increase in fossil fuel production of around 16 percent. So, on the one hand, we lecture developing countries, and on the other we do exactly what we tell others not to do. Escape routes.

Historical responsibility is also a major issue. Of course, developed countries do not want to fully acknowledge their responsibility and try all sorts of methods to calculate their GHG emissions. That is, constantly changing the reference

---

31 October to 13 November 2021, UNFCCC, 3rd Sess, UN Doc FCCC/PA/CMA/2021/10 (2022) p. 5: "36. Calls upon Parties to accelerate the development, deployment and dissemination of technologies, and the adoption of policies, to transition towards low-emission energy systems, including by rapidly scaling up the deployment of clean power generation and energy efficiency measures, including accelerating efforts towards the phasedown of unabated coal power and phaseout of inefficient fossil fuel subsidies, while providing targeted support to the poorest and most vulnerable in line with national circumstances and recognizing the need for support towards a just transition."

11. Government of Canada, *Inefficient Fossil Fuel Subsidies Government of Canada – Guidelines* (Environment and Climate Change Canada, 2023), https://www.canada.ca/en/services/environment/weather/climatechange/climate-plan/inefficient-fossil-fuel-subsidies/guidelines.html#toc5.

date for calculating the increase in GHGs and avoiding accounting from a date that would demonstrate that financial aid should be drastically increased, just like their efforts to achieve carbon neutrality.

It was also the start of a long discussion about the contributor base. The United States tried to break this North–South pattern by indicating that certain countries, such as China, should be among the contributors and not the other way round. Furthermore, during COP26, Mark Carney, then the UN Special Envoy for Climate Action and Finance, announced that the private sector is sitting on $130 trillion in liabilities that could contribute to the fight against climate change. Worse still, this amount represents only 40% of these liabilities. This indicates that a partnership between governments and the private sector is a solution to climate finance issues. Canada is joining forces with Germany to open the discussion on the contribution of the private sector, but at what price? Under what rules? Silence. In the meantime, time is running out, and on the eve of COP30 in 2025, and consequently the end of the first financial commitment by states under the Paris Agreement, few of the promised sums will have been paid out. Passively, the contributor states are demonstrating their inability to meet their commitments and the need to review the contributor base. However, the climate crisis is worsening and creating even more social injustice.

People are dying.

So, for developing countries, the feeling of betrayal is strong. This whole dynamic of diplomacy will also come under pressure with the armed conflicts in Ukraine and Palestine.

Moreover, lobbyists enter the scene during the COPs. In the wake of the Glasgow Pact, they realized that the rug could well be pulled out from under their feet. In the years following the Glasgow COP, it became clear that another game was being played behind the curtains, especially as the next three COPs would be held in oil-producing countries. The deviousness of the oil lobbyists is barely concealed: to slow down climate ambition at all costs and safeguard the industry. The ramifications are everywhere, from the electrification of transport to energy systems that would have to be added to existing ones, slowing down the principle of transition and getting a foothold in armed conflicts, some of which have energy at stake. This is also causing a considerable increase in the number of people accredited to the COPs, curious to see what is going on.

The main issue with the UNFCCC is that there is no definition of conflict of interest. If a person or entity is able to secure accreditation, they can participate in the COPs. Most lobbyists even end up on the parties' delegations. Canada grants many accreditations to the fossil fuel industry. Of course, it is not a question of arbitrating who can or cannot come, but without any notion of conflict of interest, how can we ensure that those present defend the principles of climate ambition and not their own interests? I put this question to Minister Guilbeault at a meeting of all those delegated to COP27. His vague answer never led to serious discussion.

This discussion is being led by most UNFCCC constituencies, directly with its executive secretary. Since the beginning of these discussions, a few timid steps have been taken, such as adding a series of questions on the COP registration form to disclose the funding sources of delegates. Much remains to be done, however, and several countries object to the concept of conflict of interest. Canada itself is not keen on the idea. It is often said that "if we had evidence of conflict of interest, we might be able to act." But there is proof. At COP28 and COP29, the two presidencies were at the heart of stories of conflict of interest in using their privileged positions to try to strike financial deals on fossil fuels. These facts were reported by several media outlets.[12]

---

12. Fiona Harvey, "Cop28 President Denies on Eve of Summit He Abused His Position to Sign Oil Deals," *The Guardian*, November 29, 2023, https://www.theguardian.com/environment/2023/nov/29/cop28-

14.2. The Road to Belém

So, what does the future hold for the COPs? COP30 will mark my tenth year of participation. Of course, expectations are very high. There is wear and tear in the fight against climate change. People are tired of hearing about it and are resigned to adapting, dropping the binding policies that need to be put in place to achieve carbon neutrality and prevent temperatures rising between 1.5°C and 2°C. What is more, the global chessboard changed in 2024, with more than 60 countries holding elections. The crystal ball is, to say the least, hazy on the state of the world at the end of 2025.

What has not changed, however, is that the economic crisis continues, and people are beginning to feel discouraged and angry. In any case, moralizing is no longer acceptable. Clearly, the narrative must change. Instead of telling people they will have to sacrifice and deprive themselves, the narrative should revolve around what could make them happier, healthier people. Mobilization never works by blaming those we want to mobilize. They need meaning and motivation. This is what human rights-based constituencies such as TUNGO, the Women and Gender Constituency, environmental non-governmental organizations and Indigenous Peoples organizations are doing better and better. By pooling their forces, they are increasingly becoming a counter-power, forming a bulwark against the lobbyists striving to roll back climate ambition.

However, the next few years will not be easy. The crises will intensify, and many governments will be right-wing. In Canada, resistance from Alberta and other provinces will not help meet targets. The costs of natural disasters will soar. Already, in the United States, insurance companies and major

president-denies-on-eve-of-summit-he-abused-his-position-to-sign-oil-deals; Justin Rowlatt, "COP29 Chief Exec Filmed Promoting Fossil Fuel Deals," *BBC*, November 8, 2024, *https://www.bbc.com/news/articles/crmzvdn9e18o*.

banks are pulling out. If we fail to address mitigation and the necessary move away from fossil fuel energies, the economy, and consequently people will suffer. There is a limit to adaptation, and we are already seeing it.

As far as trade unions are concerned, a future in the fight against climate change is opening. For a long time isolated, they are increasingly seen as key players for change. Why is this so? Because they represent hundreds of millions of workers, people who want to improve their living conditions and who increasingly realize that this is not compatible with the climate crisis. Canada may give tax credits to oil companies for the installation of CO_2 capture technologies, but the fact remains that floods and fires will spread year after year, affecting the dignity of those who carry the economy. Because when we talk about decarbonizing Canada, what exactly are we talking about? Just a few industries, or the entire economy? To ask the question is to answer it, and to decarbonize an economy, you need the participation of those who carry it at arm's length. Not CEOs, not prime ministers, but real people. This cannot be done without a just transition. The IPCC is clear on this point: If governments want to meet the targets of the Paris Agreement, they will require a just transition.

From a mere mention in the preamble to the Paris Agreement, the concept of just transition made its way to COP28, where a Just Transition Work Programme, recognizing labour rights as a condition, was adopted. Despite all the efforts made by trade unions and civil society on this important issue, Canada still resists putting in place a real just transition policy that would enable it to achieve its targets, without alienating its population. The stigma of the coal crisis is still very much with us.

Huge challenges lie ahead, and more than ever, the relevance of the COPs cannot be called into question.

CHAPTER 15

The UN Climate Conferences:
A Driver for "Impact Intrapreneurship"

Christophe Aura

I graduated university in international development and environmental sustainability. I volunteered at Greenpeace. I have been a public affairs consultant for years specializing in sustainable mobility, responsible finance, land protection, and clean technologies. Through those different roles, I have contributed to the creation of a protected area, to the development of the first sustainable mobility policy in Quebec, and to putting an end to fossil fuel projects.

Yet, my first participation in a Conference of the Parties (COP) of the United Nations Framework Convention on Climate Change (UNFCCC) was with ArcelorMittal, one of the world's largest steel and iron producers, in Dubai, oil state par excellence.

You think it is contradictory? I understand.

But I would say that, on the contrary, it was my most relevant experience to date in connection with COPs.

In the next few pages, I will explain the process that leads me to make this statement.

15.1. My Path to Intrapreneurship

I was introduced to the COPs while I worked at the public relations firm COPTICOM from 2017 to 2022.[1]

Of course, I had heard about them through classes and various media. I even nearly participated in the now famous 15th COP in Paris as part of a student delegation formed by the University of Sherbrooke (Quebec), before finally choosing to study for a master's degree at the University of Ottawa.[2]

But it was COPTICOM that led me to a better understanding of the COPs' workings and of their strategic utility, while I was given the mandate to produce "mission plans" for different clients, including important financial institutions in Quebec. I was responsible for identifying events, formal and informal, among the plethora that are organized at these major climate action fairs, our clients could attend to meet their strategic objectives. I also had the mandate of briefing journalists on key moments or participants to follow at the COPs, while positioning our clients in Quebec and Canadian media as climate experts.

However, as relevant as this experience was, it did not allow me to fully grasp the COPs' transformative effect for an organization and its delegates. Despite a few years of planning missions for various clients, I had never set foot in a COP.

At this time, I had always been *outside* the organizations that I accompanied, away from discussions, the reigning energy and, above all, the human exchanges and richness of the ideas that these forums facilitate.

---

1. This chapter was written while I was working for ArcelorMittal. Since then, I have returned to COPTICOM.
2. COP15 resulted in the Paris Agreement, one of the world's most significant international climate agreement ever realized. Legally binding, this international treaty ratified by 196 countries establishes the goal of limiting "the increase in the global average temperature to well below 2°C above pre-industrial levels" and to pursue efforts "to limit the temperature increase to 1.5°C above pre-industrial levels" (article 2, 1. [a]).

At least that was the case until I made a change that I never thought I would make.

15.1.1. The Transition to the Large Industry

"Hi Christophe, we are looking to make some big changes at ArcelorMittal Mining Canada. We want to become a key supplier of low-carbon iron products for steel manufacturing. I think you would be a great person to help achieve that goal. You're in?"

I was not expecting this call from a former partner, now with ArcelorMittal for some years. In fact, I admit it: At that time, I knew very little about ArcelorMittal. And what a quick online search revealed worried me. ArcelorMittal is a giant in the production of steel and iron. This implies many benefits for the economy and society in terms of generated income, jobs created, and communities supported. But also major challenges. Steel and iron production is responsible for between 8% and 10% of global greenhouse gas (GHG) emissions. Iron ore mining, the main vocation of the business unit I was about to join, by its very nature, involves managing considerable environmental risks.

So, I asked myself three questions:

- Are iron and steel essential in a world transitioning out of fossil fuels?
- Can I have a great impact on the environment and society by working at ArcelorMittal?
- Would I have unique access to levers that allow me to contribute to the transformation of an industry?

The answer to these three questions was "yes."

As a consultant, I have advised numerous clients, ranging from private companies and non-governmental organizations to public entities. But this time I had the opportunity to participate from within in the concrete implementation of projects and contribute to making lasting changes to one of the largest industries on the planet. Without knowing it at the time, I was about to forge a new identity as an "impact intrapreneur."

As an entrepreneur, the intrapreneur seeks to develop new products, practices, or processes to be deployed on the market. But as the entrepreneur develops new organizations, the intrapreneur proposes changes within an existing structure. "Impact" intrapreneurship describes any intrapreneurial action that aims to increase an organization's social or environmental performance. That is what I wanted to do at ArcelorMittal: help develop and lead projects that would make one of the world's most polluting industries a driving force for global decarbonization. As a leading iron producer and the second-largest steel company with operations across the world,[3] ArcelorMittal can generate change across an industry.

How does this all relate to COPs? They can be a fantastic intrapreneurship tool. And my participation in the 28th COP, held in Dubai in December 2023, demonstrated it.

15.1.2. The Intrapreneurial Tools Behind the Initial Guilt

To begin I want to clarify: COPs face serious limitations. From the often-timid agreements in the face of climate emergency that come out of the negotiations taking place at these events to the irony of flying thousands of people from around the world to talk about GHG emission reductions, the way in which COPs are conducted may raise questions about their legitimacy—even more so when it is held in Dubai.

I could not count how many times I heard statements like, "I feel kind of bad being here" or "How ironic is it to be in Dubai for a COP?" And I was no exception. I repeated this to myself, and started many of my conversations with these very words, as if it was our duty to do so and if our credibility

---

3. For ArcelorMittal's ranking among the steel producers: "Top Steel-Producing Companies 2024/2023," World Steel Association, June 2025, https://worldsteel.org/data/top-producers/.

depended on it, even more so in my case, now working for a large GHG-intensive industry, yet, we were all still there!

This is because COPs *do bring* real benefits in terms of climate mobilization and action. These benefits vary from one organization to another. I was in Dubai for my first concrete participation in a climate COP, so I cannot vouch for other organizations or participants. But I can explain what the two benefits are for someone like me, who sought to implement transformative projects from within a particular industry.

1) COPs can create an environment favourable to interactions that are outside the normal framework in which we operate, and that sometimes lead to conflict between different interest groups (e.g., industries vs. environmental groups). At the COP, we are all there with the same goals: to increase ambition and global capacity for climate action, and to develop new partnerships that will accelerate decarbonization efforts.

2) They can also expose various members of a same organization or company to new ideas, information, and success stories that can democratize climate knowledge and strengthen a sense of common responsibility towards the climate and environmental sustainability.

I see the first element as being firmly oriented towards external stakeholders who can help conduct an intrapreneurial project, and the second essentially oriented towards internal stakeholders. In the sections below, I detail these two elements through my experience at COP28 in Dubai.

15.2. An Environment Favourable to Improbable Connections

"Representatives from a mining company, an environmental group and a sustainable finance organization walk together into a bar…"

It could very well be the beginning of a joke, as improbable as it may seem to some. Although my COP experience did not end with a punchline at the expense of a participant, but with a moment of camaraderie. The COP was a catalyst for out-of-the-ordinary and promising meetings for more concerted climate action. However, before going into the details of these unlikely and personalized encounters, it is important to contrast them with the event that led us to them.

15.2.1. *From Excess to Closeness*

The first word that came to mind when I arrived in Dubai was: disproportion. Highways were too wide, buildings too tall—and the COP venue was too big! More than 80,000 people participated in this international event held at the Expo City Dubai. It is an immense venue built on the site for the Expo 2020 Dubai, another major international event that showcased innovation from around the world.[4]

In fact, when I arrived on site, one of the first things a colleague said as he saw me was: "You're going to go through this venue with *those* shoes?! Good luck!" He had sneakers on. I had on dress shoes. I share this little anecdote to paint a picture of the immensity of the site that was to be, for two weeks, the centre of global climate action.

The site was composed of wide outer aisles, part of which was covered with an absorbent coating normally found on running tracks, literally crossing a small village of sand-coloured buildings hosting the national pavilions of the countries represented at COP28. All my colleagues who had participated in other COPs confirmed that it was the largest they ever went to. Do not get me wrong, I was impressed by my surroundings, by the atmosphere, and I was generally excited to be physically present at my first climate COP. But my first instinct was not to tell myself that

---

4. "What We Achieved Together," Expo 2020 Dubai UAE, 2022, https://www.expo2020dubai.com/.

this place was conducive to personal encounters and human connections.

Nonetheless, Dubai has been for me the scene of enriching and personalized meetings, bringing together a variety of actors. One of the reasons for this was the context. I felt more comfortable communicating with former colleagues and partners from my past collaborations with the environmental movement who I knew were in Dubai. We were all at COP for the same reason, were we not? We could talk to each other in Canada, but it is sometimes more difficult to leave aside all the misgivings or codes that normally govern relationships between organizations such as industry and environmental groups in more "normal" settings. At home, these actors are often expected to be wary of each other.

That said, we all travelled to COP to break down barriers and strengthen collaborations. Thus, COPs allow the development of new partnerships with actors from other jurisdictions in the world (I will come back to this later in this text), but also, surprisingly, with stakeholders from home.

15.2.2. Reconnecting and Building Bridges

Following an invitation from former colleagues who were also present in Dubai, I found myself in an inner courtyard of a residential complex a little outside of the COP venue with representatives from environmental groups, academia, an employer's association and responsible finance organizations.

Quickly, the discussions became frank and simple and allowed us to address topics in a pleasant manner, which in other settings could become tense and explosive.

Questions I received included:

- Can we really produce green steel, or is it just greenwashing?
- How can ArcelorMittal Mining Canada, an iron ore mining company, contribute to decarbonization efforts?

- Is ArcelorMittal just following market trends, or is there a genuine willingness to contribute to the global decarbonization efforts?
- Is ArcelorMittal Mining Canada open to collaboration with environmental groups?

All good questions that I was happy to answer as I could. I explained what we meant by lower-carbon, or "greener," steel and how transparency standards are important to stay clear from greenwashing. I talked about ArcelorMittal Mining Canada's decarbonization plan, the concrete measures we were taking to achieve our targets, and how we could support the energy transition of the steel value chain.

I illustrated how one type of iron we produce in Canada— direct reduction iron—is a key that unlocks a decarbonization potential across the world, as it enables steel producers to transition from blast furnace technologies, very coal-dependent, to electric arc furnaces.[5]

Most of all, I reiterated my willingness, and ArcelorMittal Mining Canada's willingness, to work with all of them to achieve our common goals. And my colleagues and I intended to walk the talk.

The COP in Dubai was in December 2023. By February 2024, I was accompanying the main environmental organization in the region where ArcelorMittal Mining Canada has its operation, Quebec's Côte-Nord region, through a visit of the company's mining installations so it could see how it operates and better understand its reality in order to build stronger foundations for collaborations.

---

5.   "Direct reduction iron" is a form of iron with very few "contaminants" (e.g., silica and alumina) that can be used in electric arc furnaces: a steelmaking technology generating fewer GHG emissions than traditional blast furnaces, which are more coal-dependant. Very few places in the world can efficiently produce this kind of iron, including Canada.

At the time of writing these lines, my colleagues and I were organizing an event for COP29, held in 2024 in Azerbaijan, specifically aimed at demonstrating to stakeholders what ArcelorMittal does to reduce the iron and steel sector's GHG emissions and at presenting the challenges ahead to achieve this goal. We would thus answer some of the main questions from local stakeholders, undoubtedly also shared by other potential partners around the world, while opening the door to new collaborations.

Actions of this sort, I hope, can contribute to building bridges with organizations of all horizons, active in Canada or not. It is these bridges and the new partnerships that could come from them that will allow organizations like ArcelorMittal to think outside the box to find solutions to their environmental challenges. In this sense, I have worked in the past months with an environmental group in Quebec to get some of ArcelorMittal Mining Canada's suppliers to participate in a program through which they would be assisted in developing decarbonization plans, thereby reducing the company's indirect emissions. I have also been involved in establishing or deepening various partnerships with experts in environmental sustainability (nature-based solutions, bio-energy, tailings and residue valorization, etc.) to optimize ArcelorMittal's decarbonization and environmental policies.

This is what intrapreneurship is about. Creating new alliances to transform an organization's practices or implement new projects. External stakeholders, especially when they are not the "usual suspects," can bring innovative and refreshing ideas to persistent challenges in an organization.

As the famous saying goes, "The definition of insanity is doing the same thing over and over again and expecting a different result." Well, broadening our horizons, with all the benefits that it brings in terms of generating ideas, perhaps allows us to get wiser at the end of the day. Moreover, transparency and increased connections with stakeholders from a variety of sectors can help to correct weaknesses in projects before they are implemented and ensure that they are more acceptable for all.

The COPs create opportunities, that intrapreneurs must seize, when these diverse actors meet—whether it is on the venue's site or in the courtyard of a residential complex—away from some potential tensions that may exist at home, as long as everyone's intentions are honest and clear.

15.2.3. *The Key Is in the Intention*

Before discussing the second major contribution of COPs to impact intrapreneurship, the nature of dealings at the COPs needs to be clarified. The situation described earlier, the courtyard talks and human connections created by the COPs, may seem organic, but they are not. In reality, tensions do not all vanish when one arrives at a COP. The demonstrations against the fossil fuel industry and other large emitters that sometimes occur at and around the COPs' venue are a clear reminder of this.

So, how can you be sure that these moments of exchange and collaboration will be generated? By participating in the COP for the right reasons.

My colleague and I participated in COP28 to:

- Learn and share experiences that could accelerate the ecological transition of heavy industry.
- Explain the role of decarbonized and high-purity iron in the production of lower-carbon steel.
- Establish new collaborations rooted in a mutual willingness to help each other to positively transform business practices.

Any organization that participates in these events with the clear intention of being more mobilized and equipped to accelerate its decarbonization deserves to be there, even GHG-intensive companies. I would even say these companies *should* be there, as they are the ones who can be at the source of the biggest GHG emission reductions. They have a responsibility to act. But companies and institutions that seek to hinder climate action and ambition should not attend.

I would add that, not only is an organization's intention important, but also are the intentions of each of its participating members to the COP. Intention is fundamental to credibility. In my case, my history of collaborations with the environmental movement probably reassured some participants who would not have engaged with me otherwise, or would have, but in a very different manner. However, though my previous involvement helped build my credibility, not all industry members have experience with the environmental movement. That is okay and it certainly does not mean they do not have the legitimacy to participate in COPs.

When concrete actions are demonstrated, responsibilities are acknowledged, and participants show genuine interest, COPs can strengthen cross-sector collaboration and encourage further climate activism even within an industrial organization.[6]

This brings us to the next topic: the COPs' contribution to the "democratization" of climate mobilization within a same organization; or in other words, how COPs make accessible information that encourages climate action.

15.3. Developing a Sense of Shared Responsibility

I confess. I had already participated in a COP before going to Dubai, but a COP on biodiversity.[7]

I attended the 15th biodiversity COP, held in Montreal in December 2022, still with ArcelorMittal. Similar to climate

---

6. The fact that ArcelorMittal Mining Canada was implementing concrete climate actions, like substituting fossil fuels with bioenergy and increasing its direct-reduction iron production capacity, helped building trust with other organizations.

7. There are three "COPs." The best known is the one on climate change, which has been held every year since 1995. But there is also one on biodiversity and another on desertification.

COPs, although often more modest in size, these events focus on mobilizing delegates around actions to stop the global collapse of biodiversity.

One of the main gains from this experience was acquiring new knowledge *with my manager* on aspects that would prove critical for ArcelorMittal's own strategies, notably concerning the interdependence of climate and biodiversity issues. As ArcelorMittal Mining Canada's decarbonization strategy relies in part on forest residue-based bioenergy, biodiversity considerations cannot be overlooked.[8]

But it is the "with my manager" part that remains the most important for me. Intrapreneurship is about finding the right allies to develop projects within an organization, externally *and internally*. To build this ally community, there needs to be a common understanding of the issues at hand, especially when the intrapreneurial project consists in deep environmental transformations, which usually requires stepping outside the box. And we all need to feel this urge to have a positive impact.

In my opinion, COPs can create this sense of "common cause" among the members of a same organization. In the case of the biodiversity COP, I needed to learn more about the issues it highlighted, which I was less familiar with, and I am glad my manager was there. We developed a shared understanding of the challenges we faced, and we could now row in the same direction.

With a view to strengthen our bioenergy strategy, we started new collaborations a few months after the COP, among other things, to learn about best practices in biomass harvesting to minimize the impact on biodiversity.

This lesson I have learned from the biodiversity COP, of course, applies in full to the climate change COPs.

---

8. ArcelorMittal Mining Canada has the plan of replacing fossil fuels in its iron pelletizing plant—by far its largest source of GHG emissions—with bioenergy made from forest biomass.

15.3.1. *Reinforcing Organizational Coherence*

Going back to Dubai, we had access to many conferences, panels, workshops, and expositions from some of the best climate experts, specialized organizations, and business leaders from across the globe.

Many of those events directly or indirectly concerned ArcelorMittal's operation: events on low-carbon steel, bioenergy, the decarbonization of heavy industries, carbon pricing and markets, and more.[9]

As at the biodiversity COP in Montreal, these events were a chance for me and my colleagues to get on the same page and on the same issues and learn together about the challenges ahead as well as about the potential solutions to face them.

Months after the COP, my colleagues and I still referred to a conference that was held in Dubai during which a car manufacturer exposed its need for lower-carbon steel while exposing its limited flexibility to absorb the additional costs of this decarbonized product. This prompted us to think together more efficiently about new strategies and financial models to make our products more accessible.

This knowledge sharing accelerates climate action.

In Dubai, ArcelorMittal Mining Canada's president and CEO, vice president of corporate affairs and strategy, director of government affairs and decarbonization, and I were all present at some point and participated in some events together.

From the highest decision-making body in the organization to employees responsible for ensuring that the decarbonization measures chosen are deployed, we were all in Dubai. Organizational cohesion can only benefit from this. And organizational cohesion paves the way for faster change: an impact intrapreneur's dream. Multiple barriers related to

---

9. ArcelorMittal Mining Canada participates in Quebec's GHG emissions cap-and-trade system.

intrapreneurship arise from a lack of alignment or unequal access to key information. This leads to differentiated priorities between departments in an organization, and sometimes to a feeling that certain challenges cannot be overcome. In short, organizational incoherence can lead to resistance to change.

The information from events we attend *together* at COPs can contribute to greater organizational coherence in the future and thus to stronger intrapreneurial projects, and so does the feedback we get for the events we organize.

15.3.2. *Participating in Global Climate Capacity-Building*

COPs are a give-and-take experience. We listen to parties from around the world, but we also share our own experiences. At COP28 in Dubai, ArcelorMittal Mining Canada participated in several panel discussions to share its vision for decarbonization, projects it is undertaking, and challenges it faces.

Each participation is an opportunity to test ideas, judge the reaction of potential partners from around the world, and take in their comments. From this, projects can be adapted to be better explained within the organization and then deployed in the field.

It is in this spirit, for example, that my colleagues and I organized a panel at COP28 where we presented ArcelorMittal Mining Canada's strategy to develop a regional energy ecosystem in Quebec, where the residues of the local forest industry—by transforming them into bioenergy—could become a driving force behind the decarbonization of its iron pelleting operations. To create this event, we joined forces with a Swedish energy developer who also conceived energy ecosystems. We could then compare our experiences and learn from each other.

We also participated in an event, at the invitation of a Korean steel producer, on the steel industry scope 3 emissions (i.e., indirect emissions) where the participants, including

ArcelorMittal, had to detail and discuss their decarbonization strategy.

Again, these discussions were an opportunity to expose our strategies and ideas to other stakeholders and colleagues from around the world, in a collaborative environment. This kind of event, by exposing the strengths and improvement areas of a strategy, enables an intrapreneur to identify and seize opportunities for high-impact projects. All this amounts to one of the most critical components of impact intrapreneurship: a shared sense of responsibility towards true sustainability.

15.4. The COPs' Legacy on Impact Intrapreneurship: The Feeling of Being Part of Something Bigger Than Ourselves

As a member of the Canadian delegation, we were invited to briefing sessions with the minister of Environment and Climate Change Canada and some of his negotiators. In front of all the delegates, they presented the progress of the ongoing negotiations between the different countries to reach a new international agreement that would bring us closer to the common objective determined in the Paris Agreement.

I was not directly involved in the negotiations, but at the time I could not help feeling a sense that I was participating in something that went beyond my own personal interests or those of ArcelorMittal. All the meetings and bridges built in the last few days and all the information that we had collectively assimilated then made perfect sense. We were delegates, industry representatives, environmental groups, First Nations, youth associations, finance institutions, academics and so on at the same place, exchanging with the minister on how to create a more sustainable world.

This feeling and sense of common responsibility are the one that must drive impact intrapreneurs, and it is their duty to contribute to spreading it in their organizations.

Yes, COPs are flawed at some levels, and they need to be improved to increase their coherence with the climate imperatives, but they do foster ambition and encourage us to imagine a different world. However, to return to my initial statement, how is it that COP28, which I attended as a representative of ArcelorMittal, was the most relevant experience I have had to date with COPs?

Because I felt that I participated to reinforce this sense of "common cause" in ArcelorMittal's world by contributing to broaden its horizons and relationships and by sharing experiences with my colleagues that will strengthen our cohesion and determination in the face of the climate crisis. At COP28, I felt that I was playing my role as an impact intrapreneur.

CHAPTER 16

The Business of COP: How the Corporate Sector Took on Climate

John Stackhouse

I have been to many business conferences over the decades, from the World Economic Forum in Davos to the United Nations Population Conference in Cairo in 1994 to the World Association of Newspapers in Kyiv in 2012, but none of them compare with the UN climate change conference in Dubai, in 2024.[1] On one level, the 28th Conference of the Parties, or COP28, was similar to its 27 predecessors, in that it continued the cadence of annual meetings of countries that had signed the UN Framework Convention on Climate Change (UNFCCC), to assess their progress, share and debate policy insights, examine technological solutions, and commit to greater action—both as governments and a community of nations. On other levels, though, it was an entirely new approach to climate diplomacy, with scale and ambition from the corporate sector that both impressed and bemused a generation of COP veterans.

As was true for many there, it was one of the biggest business and investment gatherings I had seen, a kind of

---

1. The author would like to thank Taylor Neldner of Tufts University and RBC Thought Leadership for her research support.

sustainability *souk* that seemed to draw as many venture capitalists as climate activists and perhaps more portfolio managers than policymakers. I was astonished at the brash displays of climate capitalism—of new technologies for sale, of frontier investments being peddled, and of business leadership being expressed. A year earlier, at COP27 in Sharm el-Sheikh, Egypt, most of the focus was on the capital needed for developing countries suffering from the worst impacts of climate change. No one likes the word "reparations," but that was the undertone. A year before that, at COP26 in Glasgow, financial capital was centre stage, with a view that if financial institutions committed the funds, investable opportunities would follow. Dubai was different. It expressed the host city's unabashed spirit of innovation and investment—and how they could be part of the solution to climate change.

For a couple of weeks at least, "the dream in the desert," as Dubai is known, became a metaphor for a new, more business-spirited approach to climate action. It was not just the billionaire headliners—be it Bill Gates or Ray Dalio—or the corporate heavyweights like the CEOs of BlackRock, Standard Chartered, and Barclays. It was the size of the chequebooks they carried. The COP chair, Sultan Ahmed al Jaber, captured the spirit enthusiastically. He is also CEO of the Abu Dhabi National Oil Co., and was determined to make Dubai the capitalist COP. Putting his money where his mouth was, Jaber announced a new climate investment vehicle, Alterra, with $30 billion of capital behind it, including support from Brookfield, TPG, and BlackRock, three of the world's most influential institutional investors. The group went further, committing to raise $250 billion by 2050. Under Jaber's leadership, scores of other companies used the conference to add their names to global pledges on climate, from nuclear energy to methane capping. Among them, 50 oil and gas producers, representing 40 percent of global oil production, signed a charter calling for net-zero emissions by 2050 or sooner, which, to some observers, was the clearest line in the proverbial sand that the fossil fuel industry had ever drawn.[2]

Heady and heavy, and in places hedonistic, the material ambition of COP28 was more than a moment in time. It was the culmination of a decades-long journey by global business to engage in climate diplomacy and commercial and technological collaboration with the policy ambitions of their national governments. For some, this growing corporate presence at COP was an unwelcome diminishment, even abdication, of the role of government. Just ten years ago, nation-states committed to a process of national emissions targets. This has led to an expectation that governments would be held to account, even if only in the global court of public opinion—and the focus of COPs needed to be on those commitments. For others, the time had come—or was long overdue—for the private sector to play a greater role, particularly in the energy transition. Adding to that pressure, many governments had reached their fiscal capacity in the COVID-19 pandemic and ensuing economic crisis, and could no longer afford the climate bill by the time Dubai rolled around.

This shift was about more than fiscal capacity. The 2020s had ushered in a new era of technology and technological ambition, along with a new competitive dynamism between the United States and China that spanned artificial intelligence, renewable energy, and electric vehicles—the new engines of climate action. Moreover, the world was coming to grips with the growing capital needs for the energy transition—perhaps $200 trillion—and a scale of investment that only the private markets could deliver, animate, and reward. This realization was no more evident than in the United States, where in just two years the Inflation Reduction Act had deployed $100 billion of public funds and generated $500 billion in private capital for the transition—a public/private ratio that countries everywhere now hope to match.

2. "Homepage," OGDC, accessed July 23, 2025, https://www.ogdc.org/.

In this, the decisive decade, business and private capital were a new imperative, and fittingly, one of the world's great business centres, Dubai, would become their clarion call for climate capitalism.

This chapter explores the growing role of business—multinational, national, state-owned, and private—in the COP journey, and in climate process. How has that role changed over 30 years, and why? What have been the positive effects, as well as concerning outcomes? And how has business, from the Global North and South, evolved since the first Earth Summit in Rio de Janeiro, in 1992? The chapter examines not just Western corporations, including Canadian business, but also the "rise of the rest" and the role of state-owned enterprises, particularly from the Middle East and BRIC nations (Brazil, Russia, India, and China). Their interplay with global institutional investors, including large Canadian funds, as well as global supply chains is a growing force. Can these new networks help accelerate climate action where governments and democratic shareholders have fallen short? And what new forms of accountability need to be considered?

The arc of business at these UN conferences, in many ways, reflects the arc of business globally since the fall of the Berlin Wall—through financial crises, economic booms and busts, and technological revolutions. Surely, over the next three decades, the role of business in future COPs will similarly reflect future chapters of the global economy, and its fundamental role in the state of our climate. Regardless of one's views, Dubai had set a new standard for business engagement and leadership in both the COP process and, more broadly, the race to net zero. It is a trend that could shape climate diplomacy and climate action for years to come.

16.1. A Personal Journey Through COPs

My connection with UN climate diplomacy began in the early 1990s when, as a foreign correspondent for *The Globe and*

Mail based in New Delhi, I began to write about the intersection of international development and sustainability. One of my first major assignments was to profile Maurice Strong, the eminent Canadian diplomat, business executive, and international bridge builder. Strong had served as the first executive director of the UN Environment Program, in Nairobi, and had been tasked with leading the Rio Earth Summit—the conference would create the UNFCCC, which would require subsequent Conferences of the Parties, or signatories, to review and advance progress.

Strong was the glue for much of what became Rio and those subsequent COPs. His career had toggled between government (as an adviser to Pierre Trudeau), the private sector (as senior executive in the Power group of companies) and international diplomacy (as a quiet force in every environmental negotiation, and commission, through the 1970s and 1980s). He had been at the intersection of climate and commerce long before it became a focal point of UN diplomacy.

As I met up with Strong in different parts of Asia, where he had long and deep connections with all manner of governments, China most notably, he stressed to me the importance of developing countries to any future convention on climate change (a term not widely used at the time). Poverty alleviation was as important as environmental sustainability, in his view, and the fast-growing Global South would only ratchet up pressure. Strong was also aware of the importance of business to the process. He had been nominated for the UN role at Rio by Progressive Conservative Prime Minister Brian Mulroney, himself a former business executive who had already won global credibility through his advocacy for the Montréal Protocol on chlorofluorocarbon (CFC) emissions.

Globalization was not a common term back then, but the intersection of emerging markets and business—globalization by another name—was very much on Strong's mind, as a lever for environmental action. The world was also on a hinge of history. Both he and Mulroney, and the George H. W. Bush administration, had watched the collapse of the Soviet Union as a key moment for environmental progress. For one, the environmentally damaging record of communism, and state planning more broadly, had shown the need for a market-led approach, under a new era of market-minded regulations. State capital no longer worked, in their view. They also saw the rise of a corporate sector in the Global South, in countries like India and Brazil, and, of course, China, as an emerging and key opportunity.

I wrote about that transformation through the 1990s, capturing much of what I saw in my book, *Out of Poverty*, published in 2000, which took me to communities across Asia, Africa, and the Middle East, to see both the potential and limits of this new age of development. Poverty was fast collapsing in scores of countries, but the environmental and climate cost was just beginning to be seen. In my home base of New Delhi, where the population essentially doubled in the 1990s, the explosion of coal-generated power was so great I developed a tear in my lungs—something that was far more common to the millions who lived on the city's edges there. I also developed a sense of frustration with the COP process, which began formally in 1995 in Berlin (whose iconic wall represented the rise and fall of communism), as my many Indian friends, in government, media, academia, and NGOs, returned from these "confabs," as they liked to call them, with a warning that the Global North could not expect to dictate a development or growth model to the South. That frustration nearly ended the COP process after the collapse of the Kyoto Protocol in the 2000s.

I continued to connect with COPs as an editor at the *Globe and Mail*, including my time as editor-in-chief, assigning and helping shape the newspaper's coverage, particularly of Copenhagen's COP15 in 2009. It landed on the heels of the global financial crisis, with the Canadian economy still very uncertain. COP15 also became a frantic struggle to keep the process alive through an informal agreement between the world's new power rivals, the United States and China. A key aspect of the Copenhagen Accord was its pledge of $100 billion of climate-related aid to developing countries. But what Copenhagen, and the long journey of COPs, failed to recognize was the rapidly changing dynamics of the global economy—and how business (and workers) was seeing a very different world order than the one I had written about in the 1990s. For one, China was now an economic power, taking massive amounts of industrial production from the West, and millions of jobs with it. The country's growing dependence on coal was part of that transition. Meanwhile, the world had just gone through the global financial crisis, and its scar tissue had not yet healed. And then there was the explosive rise of the Internet, and social media, which was transforming industries everywhere and changing societies even more.

Business was at the heart of these multiple transformations, and in some ways, companies were becoming more powerful than governments. And yet, as I watched the COP process, it appeared to me the secondary role of business would only hold back material progress. Even in the 2010s, as disruptive technologies like artificial intelligence exploded and China grew at a breathtaking pace, the COP process seemed still focused on the North–South political tensions that had shaped the process from the early 1990s. At the time, I was settling into a post-media career at the Royal Bank of Canada, tasked with helping Canadians understand how these many disruptive forces, including climate change, were impacting the economy, communities, and clients in every sector—and to help identify what we as a country could do about it. As the COP process found a new life in Paris, in 2015, the role and expectations of business were just beginning to take off.

16.2. The Twin Journeys of COP and Commerce

To understand the evolving role of business in the COP process, we need to appreciate the evolution of both the global economy and environmental diplomacy. They are interconnected, and yet rarely are considered together. Over the past half-century, environmental diplomacy can be seen through four stages. The first began in earnest with the Stockholm Conference in 1972, which importantly focused on what was called the "human environment." A post-war Western economic order was firmly in place, rooted in multilateral agreements, most critically the General Agreement on Tariffs and Trade, and the balance sheet of leading states that helped finance the Marshall Plan, launch the World Bank, and subsidize a wide range of industries. How much of a role those states should play remained an active point of debate, and critical to the environmental movement.

That tension led to the almost simultaneous launch, one year before the Stockholm Conference, of another seminal global initiative based in Europe: the World Economic Forum (WEF). Based in Davos, Switzerland, the WEF was designed to bring business and government leaders together to discuss how trade could be liberalized and commerce advanced even faster than it had moved in the 1960s. In fact, at the time there was no clear capitalist consensus, just as there was not an environmental consensus. Many countries, Canada included, were shifting to more state intervention in economies, and even exploring variations of a "third way" between capitalism and socialism. WEF inspired many corporate leaders around the world to take on more active roles in public policy, and to increasingly see their enterprises as forces for public good, beyond products, employment, taxes, and dividends. It was a sharp shift from economist Milton Friedman's doctrine, issued in 1970, that stated "the social responsibility of business is to increase its profits."[3]

---

3. Milton Friedman, "The Social Responsibility of Business Is to Increase its Profits," *New York Times*, September 13, 1970, https://www.nytimes.com/1970/09/13/archives/a-friedman-doctrine-the-social-responsibility-of-business-is-to.html.

Into this divide waded the new multilateral environmental movement. This was no longer about 1960s-style protests against chemicals or land practices. Stockholm gave birth to a new global infrastructure to address the intersection of planet and prosperity. Curiously, business was almost absent from the Stockholm Conference, where 113 of the UN's member states and 250 NGOs shaped a new agenda—guided by a much younger Maurice Strong.

Nominated by the Trudeau Sr. government, Strong had led the UN process for two years leading up to the conference and then set out to establish the UN Environment Program, as well as a process for a new era of environmental law that would guide and govern how both states and business could act. The Stockholm Declaration was written very much in the spirit of the 1960s, to reduce consumption, combat pollution, and, in effect, place limits on growth. This was a time when Malthusian limits very much shaped global thinking, with a view the planet was at or beyond its carrying capacity. Innovation was not a dominant consideration. Instead, state-driven development, epitomized in places like Indira Gandhi's India and indeed Pierre Trudeau's Canada, was gaining pre-eminence with a view that resources needed to be managed and developed more equitably and sustainably—and a quiet assumption that business could not be trusted to do that.

The spirit of Stockholm, however, struggled to redirect global commitments through the 1970s and early 1980s, as oil shocks and stagflation made economic growth the world's overarching priority. Even the once-mighty Soviet Union could not absorb the undulations of commodity prices, and governments everywhere faced pressure from increasingly restive urban populations. In both East and West, the world had shifted from a producer-led economy to a consumer-led one, and the climate consequences were just beginning to take hold. Gutted by stagflation and soaring interest rates, fiscally challenged governments also began to see their own limits and started to sell off industrial and commercial assets, including major resource producers and emitters. The pressures were even more acute in developing countries, where the Third World debt crisis of the early 1980s led to several years of economic restructuring that curtailed state powers and put a new generation of business at the forefront. Unfortunately, the sophistication and transparency of many of those companies remained in question.

The growth of transnational corporations, largely based in the United States, Europe, and Japan, further eroded the ability of states, especially as the Reagan and Thatcher agendas elevated liberal economic thinking. As the world began to consider climate more seriously, as well as environmental challenges, global corporations were shaping economies everywhere, along with a new breed of parastatals and private corporations controlled by government insiders, from Indonesia to Pakistan and, of course, Russia and China. A new age of corporate cronyism had begun, much of it rooted in the world's heaviest emitting sectors.

In 1987, another UN effort, known as the World Commission on Environment and Development, led by former Norwegian Prime Minister Gro Harlem Brundtland, called for "a new era of economic growth." While it envisioned a stronger role for the state, and saw human development as a core principle, the imperative for economic recovery around the world continued to prevail, and so the core resolutions of the commission would not have nearly the effect they were intended to.

The Rio Earth Summit was designed to be a culmination of the Brundtland work, and a place where economic ambitions and environmental needs—as well as political necessities—could be resolved, or at least balanced. It also followed a new wave of scientific evidence of global warming, a term that took hold at Rio (Bush had just lost the 1992 US presidential election to Bill Clinton and Al Gore, in part because of his dismissiveness of this new phenomenon. In one campaign outburst, he called Gore "Ozone Man.").

But much of Rio's lofty ambitions also coincided with a second stage of climate diplomacy, under the so-called Washington Consensus. The prevailing view of international institutions, led by the International Monetary Fund and World Bank, held that development was best led by the market, and even environmental progress, with the right incentives, could be better managed by private capital. Only in hindsight did many see the tension between Rio's emphasis on state-led action and the growing view of the Washington Consensus that private capital could do more. This consensus led to the lifting of capital controls in much of the world, allowing businesses and investors to seize on opportunities in what became known as "frontier economies." But the Washington Consensus failed to anticipate the rapacious nature of this shift in countries without strong regulations and legal structures to hold companies to account; instead, private and state-tied companies in much of the former Eastern bloc and its allies extracted as much economic rent as they could, regardless of the environmental price.

A third stage of the modern climate movement began with the contemporaneous launch of both the World Trade Organization (WTO) and the World Wide Web, which towered over the global economy in the early 2000s and further constrained the power of states. The WTO allowed for the acceleration of investment and trade in the Global South, led by China, and seemed to point to a fulfilment of the Washington Consensus. At the same time, the web (and, more broadly, the digital revolution that it unleashed) allowed for a speed of commerce that, previously, had only been imagined. Both contributed, indirectly, to the global financial crisis and ensuing US recession but also changed, profoundly, the global economy—and societies—in ways that took years for COP to consider. Large corporations have changed significantly, in size, ambition and governance, through each of these stages. And yet, in hindsight, the COP process through the 1990s and 2000s failed to recognize the massive shifts underway, and further did not seize on insights from business and growing desire by many companies to play a more active role beyond their own operations.

Away from the COP process, other forces were at play. The term "ESG," for environmental, social, and governance, went mainstream in 2004, as investors and the public held companies to a new standard. It also coincided with a growing anti-corporate movement in many parts of the world, epitomized by the "Occupy" movements that erupted after the financial crisis. A new age of corporate social responsibility had taken hold and would become the basis for a fourth chapter of climate diplomacy, following the 2015 Paris summit.

COP17, in 2011 in Durban, South Africa, had been a turning point for corporate voices, as business for the first time was front and centre. So, too, was the new generation of anti-corporate voices captured in the Occupy movements. The scarring of the real economy following the financial crisis led to a greater skepticism of business, as companies were held to a new standard by workers, activists, and governments. Notwithstanding the criticism, or maybe because of it, business started to insert itself more in the COP process, and the profound economic transitions underway.

Four years later, Paris was a clear turning point for COP, not just because of its advancement of national commitments and a more performance-based approach to climate action. It also advanced the role, and responsibility, of business—at a pivotal time in commerce, investment, and trade. Paris, for the first time, called on "non-state actors" (a clinical term for business, by and large) to "scale up their climate actions." Importantly, the response from business generally was not negative; indeed, many businesses embraced the spirit of Paris, that the environment and economy were indeed interdependent.

That may not have been the case had the business cycle, and secular changes in technology, not changed the corporate mood following the global financial crisis, and ensuing recession. The crisis had put many corporations in the penalty box of public opinion. The view towards large industrial companies worsened with disasters like BP's Deepwater Horizon oil spill in 2010. Many sought to be more active in ESG matters, including climate, and saw the COP process as an important arena in which to engage governments and NGOs alike. Following the recession, at least for the firms that survived, there was also an abundance of capital to put to work.

The 2010s were increasingly defined, economically, by disinflation and low interest rates, peppered with fears of reinflation. This led to at least three seminal effects on corporate climate action. First, a global savings glut helped finance new waves of investment, including venture capital. Software was the primary beneficiary, but the abundance of capital was not lost on hardware opportunities, especially for energy technologies like hydrogen, fusion, and solar. Second, China shifted the enormous savings from its trade surplus to domestic production, with a special focus on hardware, including electric vehicles. And third, the long period of low inflation led to lower commodity prices, including for oil and gas, which encouraged many investors and energy companies to shift their portfolios to renewables.

During the 2010s, the makeup of global corporate voices also changed, led by the startling rise of Big Tech and the so-called FAANGs (Facebook, Amazon, Apple, Netflix, Google) that not only became some of the world's largest companies, worth trillions of dollars on the stock market, but also began to exert a progressive corporate voice that seemed to capture both the zeitgeist and mood of a new generation. Corporate cool had met climate, and a new age of private sector accountability for emissions—right through the supply chain—began to take root.

The pandemic did not bring this momentum to a halt, at least not initially. As societies and governments struggled under the weight of lockdowns, and economic paralysis, much of the private sector mobilized the emerging power of cloud computing and mobile technology to shift workplaces, commerce, education, and entertainment to a newly distributed economy. Work, shop, learn, and play from anywhere, anytime, became the norm, and the initial success of this economic shift empowered many employers to see their societal role anew, as champions of change. The unprecedented cooperation between government and business in the pandemic—payroll subsidies, credit facilities, Personal Protective Equipment manufacturing and distribution—furthered the belief that such public–private cooperation could be applied to climate, too.

This new spirit of corporate entrepreneurship was central to COP26 in Glasgow in 2021, which had been delayed a year by the pandemic and was quickly embraced as a gateway to a new economic thinking of "build back better." Glasgow, more than previous COPs, envisioned an economy-wide change coming out of the pandemic, with new energy, trade, and consumption systems helping to accelerate the journey to net zero. So many executives attended Glasgow that it was quickly labelled the "Business COP." It was also shaped by three business-minded governments: host Britain, which was then governed by the Conservative Party; the centrist Biden administration; and an investment-focused Chinese regime led by Xi Jinping. France's Emmanuel Macron, as the self-styled guardian of the Paris agreements, added another pro-business voice.

While Glasgow was not quite a Bretton Woods 2.0 or Marshall Plan, many government and business thinkers held a similarly spirited ambition, with state balance sheets helping to underwrite a new kind of investment boom led by the US Inflation Reduction Act. But COP26 was about more than investment. A new post-pandemic spirit saw a stronger corporate voice for regulation, especially self-regulation, as part of a broader climate ethos. More than 500 identified "fossil fuel lobbyists" (as defined by Global Witness) attended Glasgow, and arguably were important to the advancement of a global methane commitment, among other measures. Perhaps most critical to COP26 was its focus on finance. Under the leadership of billionaire media magnate Michael Bloomberg and Canadian Mark Carney, who had just stepped down as governor of the Bank of England, a powerful, private sector alliance took shape to help direct capital and markets to the enormous needs—and opportunities—of the energy transition. The Glasgow Financial Alliance for Net Zero, or GFANZ, quickly became one of the most powerful mechanisms in transition finance, with 450 banks, insurers, pension funds, and other asset managers committing to measuring their emissions, and the emissions of their clients, in order to eventually reduce them in line with national commitments.[4] The alliance was another transformation for the role of business at COP, signalling that many of the world's largest financial institutions and their collective balance sheets of $130 trillion would now be available to help the transition. To cite a common Glasgow expression, COP meant business.

16.3. The Economy, Never Far from Sight

The decades-long journey of corporations and COPs cannot be seen in isolation from the broader transformations of the global economy since the early 1990s—and the Canadian economy with it. By the time of the first COP, in 1995, a worldview of free markets had largely prevailed, and the role of governments was in decline. COP4, in Buenos Aires in 1998, included an unprecedented 150 private sector organizations. That number grew fivefold at COP6 in The Hague in 2000, where business groups helped advance a new UN Global

---

4."Homepage," Glasgow Financial Alliance for Net Zero, accessed July 23, 2025, https://www.gfanzero.com/.

Compact. By Montreal, in 2005, 1,600 private sector organizations were part of COP11.

As outlined in the previous section, the underlying economy was fast changing, fuelled by digital disruption and globalization. The World Economic Forum estimated one-fifth of emissions cuts could be achieved through digital technology. By 2010, China had also become the world's second-largest economy—and a significant investor and technology exporter. The Western approach to capitalism, though, was not embraced by all. In fact, a new chapter of globalization in the 2000s had led to a resurgence of state capitalism, led by China. The Arab state oil companies also returned to the global stage as behemoths, for investment as well as climate commitments. By 2020, 11 of the world's 15 largest state-owned enterprises were, in some form, part of COP. And they began to have a meaningful influence, especially when it came to the collective COP view of oil, gas, and coal production. According to the WEF, the top 20 state-owned enterprises in the Middle East and North Africa accounted for emissions equivalent to all of Canada, making them collectively the world's 12th biggest emitter, and they were starting to play a more significant role at COP.

If US and European multinationals were the face of global business in the 2000s and 2010s, then state-owned enterprises like Saudi Aramco, Coal India, and Sinopec would become a new voice in the 2020s at COP. And not without justification. According to the OECD, those enterprises accounted for more than one fifth of CO_2 emissions and were now critical to many countries (and the world) in meeting their emissions targets.

Canada's corporate leadership has also changed significantly through the COP years. For one, the 2000s had led to a "hollowing out" period of many Canadian multinationals, particularly in mining, as environmental leaders like Inco and Alcan were taken over by foreign companies. The Canadian economy also became much more consumption-based and services-oriented, creating a different emissions profile and, with it, strategic challenges. Another significant shift took place in the oil sector, particularly the oil sands, which ramped up investment and production through the early 2000s—and then saw a retreat of foreign investors as global oil companies reallocated their capital to lower-cost markets and new energy opportunities. That left a sizeable portion of Canadian emissions on the books of a handful of Canadian companies, like Suncor, which began to show up at COP in a more meaningful way. Often overlooked during this period was a concurrent rise of Canadian clean tech, which also developed a voice at COP. Between 2014 and 2017, Canada rose from seventh to fourth place in global clean-tech rankings, as innovators like Svante (carbon capture) and Global Energy (hydrogen) took their stories to the COP stage.

The so-called rise of the rest is not just about developing countries and emerging markets. In the COP process, it is increasingly about the national companies and sovereign wealth funds of those countries. Nowhere was this clearer than Dubai and COP28, presided over by Jaber, and his blending of economic and climate strategy. To some critics, that mix was unwise and unwelcome, especially when led by a petroleum-based economy like the United Arab Emirates. For others, it was a pragmatic approach to global capital. The climate crisis, in their view, cannot be solved without the mobilization of trillions of dollars in capital, and the advancement of many technologies that firms like the Abu Dhabi National Oil Co. (ADNOC) can be at the forefront of their development and scaling to the world.

The same tensions could be seen and heard among the Canadian delegation at COP28. For the first time, provinces—notably, Alberta, Saskatchewan and Ontario—led their own delegations, in each case with what appeared to be a larger business representation than at the federal level. Some of that commercial tone was by design: to promote economic opportunities in those provinces, and to help provincially based firms gain export markets for their clean tech, be it Ontario Power Generation's nuclear capabilities, Alberta Innovate's clean-tech portfolio, or Saskatchewan's potash resources. COP had indeed become a commercial affair, which the provincial teams had no hesitation in embracing.

In a 2024 research report entitled "Double or Trouble," the RBC Climate Action Institute estimated that about 80 percent of climate spending in Canada over the previous decade had come from government, and roughly 80 percent of that was from federal sources. Moreover, Canada's climate spending was running at less than half the pace it needed to reach for the country to get to a net zero pathway. Clearly, the private sector—investors and entrepreneurs, as well as large corporates—needs to find ways to materially increase climate-related investing.

That may continue to be one of the top challenges for future COPs, starting with COP30 in Brazil, a host country that launched the Earth Summit journey—and has a long and complex relationship between business and the environment, from cattle ranching in the Amazon basin to offshore oil development. As Brazil has seen over the past 30 years, climate progress cannot happen without business.

This view has been challenged anew, with the pro-business Trump administration withdrawing from the Paris process, and presumably assuming, at best, an observer role at COP30. The global climate movement will face a further setback if US businesses also pull back from the COP process. As a group, they are the world's leaders in technology, business models, and capital mobilization—all necessary elements of successful climate action. Whatever lies ahead for the meandering but meaningful journey of COPs, any rethinking of their model should consider a greater role for the private sector—and business from all regions—as we move into a new, and more uncertain, age of climate action and global capitalism.

CHAPTER 17

Good COP, Bad COP:
20 Years of Climate Change Negotiations

Mark Purdon

I am privileged to have been invited to contribute to this book, which gives me an opportunity to reflect on my experiences in the annual UN climate change conference—the Conference of the Parties (COP)—as well as stand back and draw some lessons for global climate governance. Since my first participation in COP11 in Montreal in 2005, by my reckoning, I have participated as an official observer in nearly every COP except for four. My most recent participation was at COP29 in Baku, Azerbaijan in November 2024. Often this has been through universities, where I have led the delegations for the Université de Montréal as well as currently that of Université du Québec à Montréal (UQAM).

What I take away from my experience is how close we as a planet were to consolidating what Bernstein described as the "compromise of liberal environmentalism"—though this now seems like something out of a dream.[1] Indeed, in nearly 20 years of COP participation, I have borne witness to an important shift in global environmental norms and international political economy. This can be described as a shift

---

1. Steven Bernstein, *The Compromise of Liberal Environmentalism* (Columbia University Press, 2001).

from liberal environmentalism that underpinned the Kyoto Protocol towards developmental environmentalism that has consolidated since the 2008 global financial crisis—a critical juncture in international political economy.[2]

Looking towards COP30, an effort to take stock of what political strategies have worked would appear to be in order. In this light, I conclude this chapter with a call towards a new approach to global climate governance that aims to take the best elements of previous sets of global norms to establish what I refer to as liberal development environmentalism.

17.1. Kyoto Dreamin'

My first time participating was in 2005 at COP11 in Montreal. Those were exciting times! I had just started a PhD in political science at the University of Toronto earlier that year. More importantly, prior to the PhD, I had worked for about a year in Cameroon, where I led one of the first consultations on opportunities for Cameroon and other countries in sub-Saharan Africa to benefit from the, then, fledgling carbon market under the Kyoto Protocol.[3] This built on master's research I had completed in 2004 in West Africa.[4]

---

2. Mark Purdon, *The Political Economy of Climate Finance Effectiveness in Developing Countries: Carbon Market, Climate Funds, and the State* (Oxford University Press, 2024).

3. Mark Purdon, *Contribution potentielle du mécanisme de développement propre (MDP) du protocole de Kyoto au développement rural du Cameroun* (Ministère de l'Environnement et de la Protection de la Nature & ACDI, 2005); Mark Purdon, *What Potential for Rural Development in Cameroon Through the Clean Development Mechanism (CDM) of the Kyoto Protocol?* (Cameroon Ministry of Environment and Nature Protection and CIDA, 2005).

4. Mark Purdon, "The Clean Development Mechanism and Community Forests in Sub-Saharan Africa: Reconsidering Kyoto's 'Moral Position' on Biocarbon Sinks in the Carbon Market," *Environment, Development and Sustainability* 12, no. 6 (2010): 1025–1050.

I remember being in Limbe, Cameroon, when, while working in one of the few internet cafes there, learning about Montréal being selected as the venue for COP11. Things seemed to be setting themselves in motion. Despite the Bush administration's refusal to implement the Kyoto Protocol by the United States (the Protocol had been signed by the previous Clinton administration but not brought before Congress for ratification), under COP rules, it managed to come into force when Russia ratified the Kyoto Protocol in late 2004.[5] In what seems like another world from today (at the time of writing in January 2025), Russia's ratification was offered in exchange for European Union (EU) support for Russia's membership into the World Trade Organization (WTO).

Meanwhile, in Canada, the Liberal Party was in power, and Stéphane Dion was minister of environment and would also become COP11 president. With the Kyoto Protocol coming into force, the meeting in Montreal also became the first session of the Conference of the Parties serving as the meeting of the Parties to the Kyoto Protocol (CMP1). It was largely anticipated at the time that Canada would become an important net buyer on the Kyoto carbon market.[6]

I was able to gain accreditation to COP11/CMP1 and went there to conduct interviews for my PhD supervisor, Steven Bernstein, while staying with friends in Montreal— my adopted hometown. But I made sure to come prepared for what promised to be an unparalleled networking experience. I actually had my CV and Cameroonian consultancy report printed, in both official languages, and in my shoulder bag at COP. It worked! I presented myself to representatives at the United Nations Development Programme (UNDP), who had just launched its Millennium Development Goals

---

5. Alexander Gusev, "Evolution of Russian Climate Policy: from the Kyoto Protocol to the Paris Agreement," *L'Europe en formation* 380, no. 2 (2016): 39–52

6. M. King, (2008). *An Overview of Carbon Markets and Emissions Trading: Lessons for Canada.* Ottawa: Bank of Canada.

(MDG) Carbon Facility there. A few months later, I received a call inviting me to intern with the UNDP in New York City during the summer of 2006. It was a fabulous start to my PhD. And, in retrospect, far too easy.

My PhD and subsequent research have focused on issues related to climate finance. This I define as transboundary flows of finance triggered by public *and* private interventions.[7] More specifically, I have considered the implementation and effectiveness of carbon markets and climate funds as different sets of climate finance instruments, which means descending from the lofty heights of the annual COP to see how climate finance instruments are being applied on the ground in developing countries and paying due attention to domestic politics. It has been a roller-coaster ride: While carbon markets were a key feature of the Kyoto Protocol, they were almost entirely absent in the decisions taken at COP15 in Copenhagen in 2009, which saw the rise of climate funds, such as the Green Climate Fund, and a commitment to providing $100 billion per year in climate finance. While carbon markets would be formally reintroduced under Article 6 of the Paris Agreement at COP21, the rules finalizing their operationalizing were only just agreed at COP29 in Baku. Indeed, COP29 has been deemed the "finance COP" for having reached decisions on carbon markets as well as, disappointingly, a commitment to increase climate finance to $300 billion per year by 2035—far greater flows are required for sustainable development and climate justice.[8] I return to these issues later in my discussion of the shift from liberal environmentalism to developmental environmentalism, after summarizing some lessons learned about the COP.

7.   Mark Purdon, *The Political Economy of Climate Finance Effectiveness.*
8.   Ariane Vartanian, "COP29 Limps Towards Agreement on Climate Finance," *Nature Reviews Materials* 10, no. 87 (2025): 1–1.

17.2. Lessons Learned About COP

Over nearly 20 years of participating in the COP, I have learned a few things about the process. First, the COP is a pre-eminent networking opportunity. If you can manage your expectations about the politics of the UN climate negotiations, as I discuss below, there are few better venues to make connections to make things happen—inside and outside the UN climate process. I mentioned my UNDP internship above, but I would be amiss if not also mentioning the NGO party held the Saturday night on the weekend between the two-week marathon negotiating sessions. (My favourite by far was the NGO party at COP15 in Copenhagen, where a multistoried bar opened up onto a full concert hall.) But for the research-minded, I suggest connecting with the research and independent non-governmental organizations (RINGOs), an officially recognized United Nations Framework Convention on Climate Change (UNFCCC) constituency.[9] They host regular morning meetings that allow one to catch up on the negotiations, as well as allowing for delegates to actively participate in their various thematic groups.

Second, the COP is also relatively accessible. Of course, more can be done to level the playing field between developed and developing countries, as well as civil society and the private sector, but the UNFCCC is much more open and transparent than other major international forums, like the WTO[10] or the annual meeting of the World Economic Forum in Davos.[11]

---

9. "Homepage," Research and Independent Non-Governmental Organizations, accessed July 4, 2025, https://ringosnet.wordpress.com/.

10. Padideh Ala'i and Katayoon Beshkardana, "The Limits of Transparency: China, the United States and the World Trade Organization," in *Cultures of Transparency*, eds. Stefan Berger, Susanne Fengler, Dimitrij Owetschkin, and Julia Sittmann (Routledge, 2021), 135–153; Steve Charnovitz, "Transparency and Participation in the World Trade Organization" *Rutgers Law Rev* 56, no. 4 (2005): 927.

11. Desmond McNeill, "The World Economic Forum: An Unaccountable Force in Global Health Governance?" *Global Policy* 14, no. 5 (2023): 782–789.

For example, I have been able to organize at least three official side events to the UNFCCC over the years with virtually no resources: the UNFCCC allocates official side events a conference room, complete with IT services, while the side event is announced in official COP bulletins. The most memorable side event I organized was at COP19 in Warsaw in 2013, where a PhD colleague and I were able to orchestrate a panel on the California–Quebec carbon market with colleagues in California. The side event and subsequent dinner were the first, high-level meeting between the Quebec minister of environment (Yves-François Blanchet) and California's secretary for environmental protection (Matthew Rodriquez). I did that while a post-doctoral candidate at the London School of Economics—that is, without almost any resources.

Similarly, the Institut québécois du carbone (IQCarbone), a climate policy research institute that I used to lead, was able to secure a kiosk at COP21 in Paris for a reasonable fee. We found our kiosk placed next to those of other major organizations in COP21's Green Zone. It should be stressed that, since Paris, all COPs have divided the grounds into a Green Zone, similar to a trade fair and open to the public, and a Blue Zone, where the negotiations happen and only open to official parties and observers. The two zones abut one another, which allows for an important exchange of ideas.

Third, the UN process is not the only, nor even the best, venue for climate change politics. Indeed, there is an extensive critique of the UNFCCC as a political institution for solving one of the world's most pressing problems.[12] When participating in COP, one should bear in mind that the

12. Harro Asselt and Fariborz Zelli, "Connect the Dots: Managing the Fragmentation of Global Climate Governance," *Environmental Economics and Policy Studies* 16, no. 2 (2013): 137–155; Rakhyun E. Kim, "Is Global Governance Fragmented, Polycentric, or Complex? The State of the Art of the Network Approach," *International Studies Review* 22, no. 4 (2020): 903–931; David Victor, *Global Warming Gridlock* (Cambridge University Press, 2011).

UNFCCC is quite a weak international institution, without much enforcement power. The COP also works on a consensus basis, meaning that decisions taken at COP reflect broad but shallow agreement among states that are party to the UNFCCC. On occasion, individual countries have been able to derail the decision-making process—most spectacularly at COP15 in Copenhagen in 2009, where the international community failed to negotiate a meaningful extension of the Kyoto Protocol.[13]

But this should not be a cause for despair given that so much is happening outside the UN climate change regime—in what many have referred to as the climate change "regime complex."[14] Indeed, just down the road from the COP15 venue, at a luxury hotel in Copenhagen, a group of transnational NGOs, multilateral firms, and other non-state actors were busy crafting rules for the voluntary carbon market that is independent from the UNFCCC.[15] Similarly, the transnational carbon market linking California and Quebec has taken place under the auspices of the Western Climate Initiative (WCI) independent of the UNFCCC.[16] Policy actors can

13. Daniel Bodansky, "The Copenhagen Climate Change Conference: A Postmortem," *American Journal of International Law* 104, no. 2 (2010): 230–240; Radoslav S. Dimitrov, "Inside Copenhagen: The State of Climate Governance," *Global Environmental Politics* 10, no. 2 (2010): 18–24.

14. Kenneth W Abbott, "The Transnational Regime Complex for Climate Change," *Environment and Planning C: Government and Policy* 30, no. 4 (2012): 571–590; Robert Keohane and David Victor, "The Regime Complex for Climate Change," *Perspectives on Politics* 9, no. 1 (2011): 7–23.

15. Steven Bernstein et al., "A Tale of Two Copenhagens: Carbon Markets and Climate Governance," *Millennium: Journal of International Studies* 39, no. 1 (2010): 161–173.

16. David Houle, Erick Lachapelle, and Mark Purdon, "The Comparative Politics of Sub-Federal Cap-and-Trade: Implementing the Western Climate Initiative," *Global Environmental Politics* 15, no. 3 (2015): 49–73; François Roch and Jacques Papy, "L'Entente de liaison des marchés du carbone de la Western Climate Initiative: enjeux insti-

also act unilaterally and have a major global impact on climate issues, perhaps best exemplified by the adoption of the Inflation Reduction Act by the United States in 2022.[17] Offering hundreds of billions of dollars in clean technology subsidies, the Inflation Reduction Act is a leading example of a surge of interest in green industrial policy in recent years, albeit an issue only marginally discussed by the UNFCCC.[18]

Another important issue on which the COPs are largely silent is international trade, especially given growing interest in using border carbon adjustments to catalyze climate clubs.[19] The governance of international tariffs is more the domain of the WTO, though it is also important to acknowledge that the WTO's authority is being sapped by the quiet withdrawal of support by the United States.[20] For example, the prospect of a climate club became much more real when the European Union reached a provisional agreement in late 2022 to introduce a Carbon Border Adjustment Mechanism (CBAM). While CBAM has been designed to conform to the rules of the WTO, there is a concern that the CBAM might allow the European Union to engage in protectionism in the guise of climate action.[21] Despite their growing prominence,

tutionnels et juridiques pour le Québec," *Revue générale de droit* 49, no. 1 (2019): 67–109.

17. John Larsen et al., *A Turning Point for US Climate Progress: Assessing the Climate and Clean Energy Provisions in the Inflation Reduction Act* (Rhodium Group, 2022).

18. Bentley Allan, Joanna I. Lewis, and Thomas Oatley, "Green Industrial Policy and the Global Transformation of Climate Politics," *Global Environmental Politics* 21, no. 4 (2021): 1–19.

19. Indra Overland and Mirza Sadaqat Huda, "Climate Clubs and Carbon Border Adjustments: A Review," *Environmental Research Letters* 17, no. 9 (2022): 093005.

20. Alan Wolff, "Is the World Trade Organization Still Relevant?" *Peterson Institute for International Economics Policy Brief* 24, no. 15 (2024): 1–13.

21. UNCTAD, *A European Union Carbon Border Adjustment Mechanism: Implications for Developing Countries* (Geneva, 2021).

climate clubs and international trade policy in general do not figure prominently in the COP negotiations.

A final lesson learned is that there continues to be insufficient understanding about how decisions made at the COP are implemented on the ground and how effective they are in reducing emissions and contributing to decarbonization. Too often, the belief that the COP is where all climate politics happens has, regrettably, focused research attention at the global level rather than considering how climate policy is being implemented and how effective it is. This is problematic as debates about policy effectiveness at the global level—especially at the COPs themselves—are not necessarily informed by experience on the ground or empirical research. As I have long suggested, there is a real need for greater research into comparative climate change politics, particularly into thorny issues of policy effectiveness, in order to inform the design of climate finance instruments and other tools of climate change cooperation.[22]

Consider the debate on international carbon markets and climate funds. Many critics of international carbon markets suggest that they be replaced by projects and programs supported through official development assistance (ODA) and implemented in partnership with international development agencies like the UNDP and World Bank. In a recent book, I demonstrate that, despite the shift from carbon market towards climate funds supported by public finance, the actual implementation and effectiveness of different climate finance instruments backed by these different strategies in Tanzania, Uganda, and Moldova did not appreciably change over the period 2008 to 2018.[23] Climate finance instruments were consistently more effective in Uganda and Moldova

22. Mark Purdon, "Advancing Comparative Climate Change Politics: Theory & Method," *Global Environmental Politics* 15, no. 3 (2015): 1–26.

23. Mark Purdon, *The Political Economy of Climate Finance Effectiveness*.

than Tanzania, despite differences in state capacity between sub-Saharan Africa and the former Soviet Union. But it is not difficult to find strong opinions for or against carbon markets or climate funds amongst participants of the annual UN climate change conferences.

17.3. The Shift from Liberal Environmentalism to Developmental Environmentalism

In addition to lessons learned about the COP, over nearly 20 years of UN climate negotiations, I have borne witness to an important change in norms governing climate change and other international environmental issues.

As suggested earlier, when I first began participating in the UN climate process in 2005, the climate change regime adhered to what Bernstein has described as a global environmental norm of *liberal environmentalism*.[24] Indeed, the UNFCCC's effort to establish a global carbon market under the Kyoto Protocol was consonant with neoliberal ideas, such as private property, free trade, and self-regulating markets. In particular, the Clean Development Mechanism (CDM) of the Kyoto Protocol came to embody liberal environmentalism in terms of its reliance on the private sector, emphasis on results in the form of verifiable emission reductions, and the limited role accorded the state for implementation.

But, as is widely known, the Kyoto Protocol's carbon market collapsed after negotiations at the 2009 UN climate change conference in Copenhagen failed to secure a meaningful extension of the Kyoto carbon market approach. In my recent book, I argue that the breakdown of liberal environmentalism is consistent with broader changes in global political economy since the 2008 financial crisis—a watershed moment in international political economy.[25] As Birdsall and Fukuyama argued, "[i]f the global financial crisis put any

24.   Steven Bernstein, *The Compromise of Liberal Environmentalism*.
25.   Mark Purdon, *The Political Economy of Climate Finance Effectiveness*.

development model on trial it was the free market or neoliberal one, which emphasized a small state, deregulation, private ownership, and low taxes."[26] Liberal environmentalism has been superseded by two new sets of global norms operating in parallel: developmental environmentalism and transnational climate governance.

Before exploring these two new global environmental norms in more detail, it is helpful to reflect on the timing of the 2008 global financial crisis relative to COP15. In the lead up to COP15, interest in carbon markets was still high. A second phase of the European Union's Emissions Trading System had been launched in 2008 and demonstrated robust prices.[27] Similarly, the US federal government had come very close to adopting comprehensive cap-and-trade legislation ahead of Copenhagen (for an introductory overview of cap-and-trade and other elements of carbon markets, see my article in *The Conversation Canada*[28]). President Barack Obama won the 2008 US presidential election with a Democratic majority in Congress and was poised to act on his promise to introduce such a system.[29]

Nonetheless, the 2008 global financial crisis ultimately undermined climate efforts. For example, cap-and-trade

---

26. Nancy Birdsall and Francis Fukuyama, "The Post-Washington Consensus: Development After the Crisis," *Foreign Affairs* 90, no. 2 (2011): 45–46.

27. Germà Bel and Stephan Joseph, "Emission Abatement: Untangling the Impacts of the EU ETS and the Economic Crisis," *Energy Economics* 49 (2015): 531–539.

28. Mark Purdon, "COP29: Canada Needs to Start a Real Conversation About International Carbon Markets," *The Conversation Canada*, November 14, 2024, https://theconversation.com/cop29-canada-needs-to-start-a-real-conversation-about-international-carbon-markets-242152.

29. Jon Birger Skjærseth, Guri Bang, and Miranda A Schreurs, "Explaining Growing Climate Policy Differences Between the European Union and the United States," *Global Environmental Politics* 13, no. 4 (2013): 61–80.

legislation in the United States never made it to the US Senate. One reason for its defeat was that public support for climate policy collapsed following the financial crisis.[30] In turn, the failure to pass major cap-and-trade legislation deprived Obama of a key bartering chip as he arrived in Copenhagen in late 2009.[31] While certainly many factors contributed to the failure in Copenhagen, "the global financial crisis sucked political attention, energy and momentum from the climate issue."[32] As Meckling and Allan demonstrate, alternatives to liberal environmentalism grew more prominent in global climate policy discourse after the global financial crisis.[33]

A first alternative to liberal environmentalism to emerge was transnational climate governance, interest into which grew considerably since Copenhagen.[34] The idea here was that the UN climate regime was too rigid and top-down. It frustrated climate action being undertaken by a broader variety of actors at multiple level of governance from the bottom-up. The hope was that even modest actions would be incremental actions, and would accumulate over time to deliver transformational change.[35] However, some import-

---

30. Lyle Scruggs and Salil Benegal, "Declining Public Concern About Climate Change: Can We Blame the Great Recession?" *Global Environmental Change* 22, no. 2 (2012): 505–515.

31. Wang Mou, Pan Jiahua, and Chen Ying, "Analysis on American Clean Energy and Security Act," *Advances in Climate Change Research* 6, no. 4 (2010): 307–312.

32. Peter Christoff, "Cold Climate in Copenhagen: China and the United States at COP15," *Environmental Politics* 19, no. 4 (2010): 637–656.

33. Jonas Meckling and Bentley B. Allan, "The Evolution of Ideas in Global Climate Policy," *Nature Climate Change* 10, no. 5 (2020): 434–438.

34. Harriet Bulkeley et al., *Transnational Climate Change Governance* (Cambridge University Press, 2014); Andrew Jordan et al., eds, *Governing Climate Change: Polycentricity in Action?* (Cambridge University Press, 2018).

35. Kelly Levin et al., "Overcoming the Tragedy of Super Wicked Problems: Constraining Our Future Selves to Ameliorate Global Climate Change," *Policy Sciences* 45, no. 2 (2012): 123–152.

ant questions might be raised about transnational climate governance, which, with its faith in non-state actors, bears an uncanny similarity to liberal environmentalism. Furthermore, research into transnational climate governance has been overwhelmingly focused on potential impacts and not ex post efforts to gauge policy effectiveness in terms of reducing emissions and contributing to decarbonization.[36] Transnational climate governance might be the logical extension of liberal environmentalism rather than a break from it.

The delegitimization of neoliberal economic ideas following the 2008 global financial crisis allowed a much different set of ideas, what I describe as developmental environmentalism, to gain traction at COP15. It is true that carbon markets were kept alive at COP15 through the Ad Hoc Working Group on Long-term Cooperative Action (AWG-LCA) process but carbon markets lacked momentum as demand for carbon credits and allowances via the Kyoto Protocol evaporated. The turn towards green industrial policy, particularly the adoption of the US Inflation Reduction Act in 2022, is perhaps the culmination of a major shift in thinking about economy and environment that can best be characterized as developmental environmentalism.

When it comes to climate finance, in contrast to liberal environmentalism, developmental environmentalism is characterized by (i) greater reliance on public donor financing for international climate funds, (ii) the softening of evaluation frameworks for demonstrating policy effectiveness, and (iii) a stronger role for the state in policy implementation. I briefly discuss these three characteristics below, though the interested reader might consider my recent book for more

---

36. Thomas N. Hale et al., "Sub- and Non-State Climate Action: A Framework to Assess Progress, Implementation and Impact," *Climate Policy* 21, no. 3 (2021): 406–420.

detail.[37] In terms of climate finance for developing countries, COP15 was a watershed when committing to:

> Scaled up, new and additional, predictable and adequate funding [and] improved access shall be provided to developing countries […] In the context of meaningful mitigation actions and transparency on implementation, developed countries commit to a goal of mobilizing jointly US$100 billion dollars a year by 2020 to address the needs of developing countries. This funding will come from a wide variety of sources, public and private, bilateral and multilateral, including alternative sources of finance.[38]

The emphasis on climate funds would dovetail nicely with the use of stimulus funding to restart the world economy in the wake of the 2008 global financial crisis. The move towards climate funds was reinforced in Copenhagen through agreement to establish a Green Climate Fund through which a "significant portion of [the US$100 billion a year by 2020 in] funding should flow."[39]

While there was a delay, the Organisation for Economic Co-operation and Development (OECD) has estimated that $100 billion goal was achieved in 2022.[40] But nearly 80 percent of this goal has been based on the provision of ODA. Questions remain if such funding is really "new and additional" and not simply the relabelling of existing ODA contributions. For example, climate ODA represented nearly half of all ODA provided by the OECD in 2022, though it really only first appeared in 2010, as shown in Figure 17.1. Similarly, the onset of the war in Ukraine

37.   Mark Purdon, *The Political Economy of Climate Finance Effectiveness*.
38.   *Copenhagen Accord*, UNFCCC, 15th Sess, Un Doc FCCC/CP/2009/11/Add.1 (2009) Dec 2/CP.15 at para 8.
39.   *Copenhagen Accord*, UNFCCC, 15th Sess, Un Doc FCCC/CP/2009/11/Add.1 (2009) Dec 2/CP.15 at para 8.
40.   OECD, *Climate Finance Provided and Mobilised by Developed Countries in 2013-22: Climate Finance and the USD 100 Billion Goal* (OECD, 2024).

is likely to have diminished donor financing for international development since 2022.[41] Overly reliant on public donor financing, such as ODA, there are concerns about levels of climate finance required for decarbonization and adaptation, despite all institutional innovations introduced over the years through various COPs. The need for international climate finance in developing countries is only growing: the main decision of COP29 was to increase the climate finance goal to $300 billion by 2035, a figure that many experts believe is too low.[42]

Figure 17.1. ODA for climate and non-climate purposes, according to data from OECD Development Assistance Committee (constant 2022 USD).

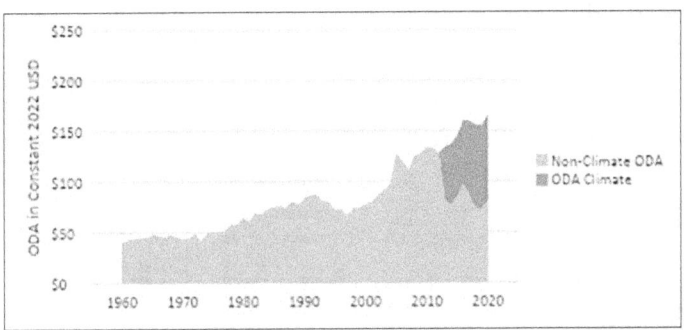

Sources: OECD Data Explorer, DAC1: Flows by donor (ODA+OOF+-Private).[43] Note that only "net disbursements" of ODA in constant 2022 USD were retained (officially referred to as "I. Official Development Assistance (ODA) (I.A + I.B)"). Climate Finance and the USD$100 billion goal, OECD.[44] Data presented by OECD are in current USD and were converted into constant 2022 for the purposes of analysis.

41. Carina Strøm-Sedgwick and Pinar Tank, "The Impact of the Ukraine War on Donor Priorities," *PRIO Policy Brief* 17 (2022): 1–4.
42. Amar Bhattacharya et al., *Raising Ambition and Accelerating Delivery of Climate Finance: Third Report of the Independent High-Level Expert Group on Climate Finance (IHLEG)* (Grantham Research Institute on Climate Change and the Environment, LSE, 2024).
43. https://data-explorer.oecd.org, OECD, accessed January 30, 2025.
44. "Climate Finance and the USD 100 Billion Goal," OECD, accessed January 30, 2025, https://www.oecd.org/en/topics/sub-issues/climate-finance-and-the-usd-100-billion-goal.html.

17.4. Towards Liberal Developmental Environmentalism?

Overall, there has been an important shift in global environmental norms from liberal environmentalism to developmental environmentalism over the past 20 years of UN climate change negotiations. Neither has been successful in delivering the transformational change required. In light of the performance of the UN climate change regime to meaningfully engage developing countries on climate change mitigation and adaptation, I submit that it might be fruitful to combine elements of liberal environmentalism and developmental environmentalism. I describe this as liberal developmental environmentalism.[45] See Table 17.1 for its defining characteristics relative to global environmental norms discussed.

Table 17.1. Characteristics of Liberal Environmentalism, Developmental Environmentalism and Liberal Developmental Environmentalism

	Global Environmental Norm		
	Liberal Environmentalism	**Developmental Environmentalism**	**Liberal Developmental Environmentalism**
Principle source of financing	Private sector	Public sector	Public and private sectors at funding levels to (prudently) defy comparative advantage
Role of the state	Limited role	Essential partner	Constructive roles between state and non-state actors
Evaluation frame	Rigorous results-based approach; anticipates need for carbon market fungibility	More traditional approach to development cooperation; independent of carbon markets	Rigorous results-based approach; anticipates need for carbon market fungibility

Source: Purdon (2024).[46]

---

45. Mark Purdon, *The Political Economy of Climate Finance Effectiveness*.
46. Purdon, *Climate Finance Effectiveness*.

Most importantly, liberal developmental environmentalism would balance the needs of the international community for the demonstration of policy effectiveness with developing countries' need for development. In practical terms, liberal developmental environmentalism would prioritize climate mitigation actions that states are already pursuing but also include a more robust evaluation framework to attract resources from an international community that still demands results. Such an approach might go further to attract private finance for climate mitigation efforts in developing countries and avoid current dependency on public financing. Furthermore, instead of privileging either state or non-state actors, this would cultivate synergies between the two in the transition to low-carbon development. I elaborate on these dimensions of liberal developmental environmentalism below, as they might be envisioned for tackling climate change mitigation.

The first would be to establish an evaluation framework that responds to the preferences of transnational private actors, international donors, and central governments hosting climate finance initiatives. To appeal to private sector actors and donors, climate finance instruments would create more rigorous evaluation frameworks that offer greater confidence that interventions supported by climate finance are leading to emissions reductions. However, the developmental dimension of climate finance would be promoted by ensuring that the prices for emission reductions enable a country to defy its comparative advantage, realize its developmental potential, and achieve transformational change.

However, any climate finance instrument based on liberal developmental environmentalism needs to avoid setting the prices of emission reductions only on neoliberal principles, such as the opportunity costs associated with the current comparative advantages of developing countries. Instead, carbon prices should be based on the costs of defying (perhaps prudently) existing comparative advantage. The methodological challenges implied in realistically endeavouring to transform a country's comparative advantage are rather uncharted territory and would benefit from more sophisticated economic analysis. But it is also worth considering how fabulously low carbon prices associated with both carbon market and climate fund initiatives have been: ranging from US$2.5 to US$12.5 per tonnes of carbon dioxide equivalent (tCO2e) in research I conducted over 2008–2018.[47] Such prices are well below the social cost of carbon as well as prices currently observed on major carbon markets. In this respect, a price floor starting at least US$15 per tCO2e and rising with inflation could be considered as a point of departure for discussion of a liberal developmental environmental global price for carbon. Notably, such carbon price floors serve to stabilize prices in other emission trading systems, such as that of the California–Quebec carbon market.[48] Higher carbon prices might also be expected to allay the regulatory burden, as they would allow effective climate actions to be more easily distinguished from counterfactual business-as-usual scenarios.

17.5 Conclusion

By the time COP30 rolls around in Belém, Brazil, I will have participated in nearly every COP over the past 20 years. That is a lot of air miles and emissions. While I buy carbon offsets, I have also learned a few things along the way that I believe justifies continued engagement in the COP process. My research is essentially focused on connecting the global and domestic policy levels, which has given me an interesting perspective on what the COPs can and cannot do. Using an interdisciplinary approach that integrates international relations, comparative politics, and policy studies, I have also

---

47. Purdon, *Climate Finance Effectiveness.*
48. Mark Purdon et al., "Climate and Transportation Policy Sequencing in California and Quebec," *Review of Policy Research* 38, no. 5 (2021): 596–630.

been keenly interested in research policy effectiveness in countries (including Canada and Quebec). This has led me to question prevailing global environmental norms of liberal environmentalism (and its extension, transnational climate governance) as well as developmental environmentalism, which has gained greater traction since the global financial crisis. My hope is that an integration of the best elements of previous approaches to global climate governance might be progressively brought together in what I refer to as liberal developmental environmentalism. It is at least a great research question to motivate another generation of investigation into global climate governance.

Authors: Lauren Touchant, Elie Klee, Mikayla Carson, Jessica Lee, Sarah Mei Lyana, Kate Mooney, Henry Sipos, and Thomas Burelli

COP Number	City, Country	Attendance (Parties + Observer States + Observer Organizations)	Size of the Canadian Delegation (Number of Participants/ Total of Party Participants)	Highlights of the COP	Canada's Contribution	Outputs/Outcomes of the COP
COP1 (1995)	Berlin, Germany	117 states, comprising 757 participants 53 observer states, comprising 112 participants 196 observer organizations,[1] comprising 1,056 participants 556 media, comprising 2,044 participants	25/757	• Discussions about the permanent Secretariat location. • Discussions on technology transfers between developed countries (Annex I) and developing countries and other states (Annex II). • Discussions about joint implementation (JI) to create a carbon market between developed countries and developing countries. • Discussions about the adequacy of commitments, including the common but differentiated principle defended by developing countries and the level of responsibility of developed countries. • Discussions about GHG targets: Developed countries should set higher targets; however, they argue that targets should instead be based on levels of economic development. • Island states declare themselves as "the Earth's early warning system" for climate change. • Discussions about the governance structure and funding of the UNFCCC and the Conference of the Parties.	• Canada volunteered to host the Secretariat in Toronto, but Canada suggested that the UNFCCC should be located on neutral grounds. • Canada supported a JI pilot project as well as technological transfer between developed and developing countries. • Canada financially committed $1 million for five years to the Secretariat budget.	• Permanent location of the UNFCCC is Bonn, Germany. • Agreement on the framework for future COP. • Parties agree on the JI pilot project and good intake from developing countries. • Funding priorities allocated to developing countries. • Rules of procedures moved to COP2. • Green Environmental Fund to serve as the interim financial mechanism.

1. United Nations Secretariat units and bodies; United Nations specialized agencies and related organizations; intergovernmental organizations and non-governmental organizations. The non-governmental organizations are always the largest category among the observer organizations.

COP Number	City, Country	Attendance (Parties + Observer States + Observer Organizations)	Size of the Canadian Delegation (Number of Participants/ Total of Party Participants)	Highlights of the COP	Canada's Contribution	Outputs/Outcomes of the COP

COP2 (1996)	Geneva, Switzerland	130 states, comprising 500 participants
		
		6 observer states, comprising 17 participants
		
		103 observer organizations, comprising 339 participants
			20/500	• Top 15 countries responsible for 55% of GHG emissions report an increase of their emissions (exceeding 1990 levels).
				
				• The Second Assessment Report is the highlight of COP2. It highlighted the scientific consensus on human-induced climate change and its potential impacts. COP2 was influenced by the findings of SAR, particularly regarding the need for climate action to mitigation climate change and the associated risks.
					• Canada recognized the scientific basis of the Second Assessment Report for urgent climate action.
					
					• Canada emphasized the importance of increasing cooperation to address climate change, including the involvement of industry and NGOs. In the Committee on Development and Transfer of Technology, Canada advocated for the "creation of an environment enabling input from the private sector" through a business consultative mechanism that would not be open to NGOs.
					
					• Announcements that the IPCC will be holding a workshop on adaptation measures in Canada.
						• COP2 ended with strong political signals toward strengthening the commitments on the part of industrialized countries to reduce GHG emissions beyond the year 2000.
						
						• Accentuation of disagreements:
						  ○ Balance between an agreed set or menu of policies and measures;
						  ○ Measures to offset the economic losses and costs of climate actions for developing countries and oil-exporting states.
						
						• Decision to produce a national inventory of anthropogenic emissions by sources and removals by sinks of all GHGs as well as a general description of steps taken to implement the FCCC.
						
						• Rejection of language that commits non-Annex I Parties to fund mitigation projects after developing countries accused Annex I Parties of shifting the burden of implementation.
						
						• While several delegations endorsed SAR, several states, including oil-producing states and Australia, did not want to make recommendations.
						
						• The declaration recognized the urgency of climate action, but it failed to call for binding measures and did not establish a GHG reduction commitment below 1990 levels.

COP Number	City, Country	Attendance (Parties + Observer States + Observer Organizations)	Size of the Canadian Delegation (Number of Participants/ Total of Party Participants)	Highlights of the COP	Canada's Contribution	Outputs/Outcomes of the COP
COP3 (1997)	Kyoto, Japan	155 states, comprising 1,534 participants 6 observer states, comprising 29 participants 278 observer organizations, comprising 3,865 participants 483 media, comprising 3,712 participants	51/1,534	• Negotiation of the first legally binding agreement. • The United States played a significant role in shaping the institutional structure of emissions trading. • The European Union provided the ambition to establish the Kyoto targets; it was supported by developing countries. • G77/China group defeating New Zealand's proposal to obtain voluntary commitments for developing countries. • Brazil brokered the Clean Development Fund, which, with the support of the United States, evolved into the Clean Development Mechanism. The proposal survived despite resistance from China and India and was reaffirmed through the voluntary participation of Mexico and the Republic of Korea.	• Canada opposed compensation fund proposed by Iran. • Canada supported New Zealand's proposal to seek commitments from developing countries to adopt binding emissions limitation targets in a third commitment period.	• Adoption of the Kyoto Protocol, with Annex I Parties agreeing to reducing their overall emissions of six GHGs by at least 5% below 1990 levels between 2008 and 2012. • The Protocol establishes emission trading, joint implementation between developed countries, and a clean development mechanism to encourage joint emissions reduction projects between developed and developing countries.
COP4 (1998)	Buenos Aires, Argentina	126 states, comprising 609 participants 6 observer states, comprising 21 participants 130 observer organizations, comprising 572 participants 299 media, comprising 883 participants	43/609	• G77/China was fractured by a decision to establish the Berlin Mandate, a process for negotiating stronger commitments from developed countries to reduce GHGs. • Argentina tried to place voluntary commitments for developing countries on the agenda, which prompted US action to sign the protocol. • G77/China highlights the poor performances of Annex I Parties	• Canada, supported by Australia proposed that the private sector should be the main vehicle for technology transfer. • Canada and South Korea supported work on inventories for sources of new technologies and gap identification. • Canada wanted the private sector to be a vehicle for technology transfer.	• Decision of the United States to sign the Kyoto Protocol. • Adoption of the Buenos Aires Action Plan establishing a deadline for finalizing work on the Kyoto mechanisms (joint implementation, emissions trading, and the clean development mechanism). • Did not finalize many crucial details.

COP Number	City, Country	Attendance (Parties + Observer States + Observer Organizations)	Size of the Canadian Delegation (Number of Participants/ Total of Party Participants)	Highlights of the COP	Canada's Contribution	Outputs/Outcomes of the COP
COP5 (1999)	Bonn, Germany	138 states, comprising 639 participants 3 observer states, comprising 13 participants 136 observer organizations, comprising 626 participants 30 media, comprising 68 participants	40/639	• The Conference worked on the rules for the clean development, joint implementation, and emissions trading mechanisms. • Discussions about how carbon sinks (such as forests) could be used to offset emissions. • Progressed on the development of the clean development mechanisms. • Recognition of the need to address the impacts of climate change. There were discussions about the creation of an adaptation fund. • US withdrawal from the Kyoto Protocol, which led to discussions about how to proceed with the Protocol without the participation of the United States. • Developing countries continued to emphasize the need for financial support and technology transfer.	• Canada supported the development of a set of guidelines for the creation of national registries for each mechanism. • Canada opposed the application of the principle of common but differentiated responsibilities on compliance. • Canada opposed a triggering role for the Secretariat, reaffirming that it should play a simple role of information gathering. • Canada supported the appointment of a special facilitator to assist negotiations and help Parties realize the BAPA by COP6.	• No significant breakthrough but progress on the operationalization of the mechanisms.

COP Number	City, Country	Attendance (Parties + Observer States + Observer Organizations)	Size of the Canadian Delegation (Number of Participants/ Total of Party Participants)	Highlights of the COP	Canada's Contribution	Outputs/Outcomes of the COP
COP6 (2000)	The Hague, Netherlands	177 states, comprising 2,195 participants 4 observer states, comprising 20 participants 323 observer organizations, comprising 3,835 participants 443 media, comprising 944 participants	81/2,495	• Discussion of "cluster" issues, including technology transfer, adverse effects, and guidance to Global Environment Facility (GEF). • Debate over how much carbon sequestration should count toward emission reduction. • Discussion on emissions trading, JI and CDM. • Countries with economies in transition (EITs) and developing countries focused on financial and technical support to build capacity for climate mitigation and adaptation. • Debate over whether to have one decision covering both UNFCCC Article 4.8/4.9 (adverse effects, LDCs support) and Kyoto Protocol Article 3.14 (impact of response measures) or two separate decisions. • Disagreements over legally binding nature of reduction in excess emissions from country's next commitment period target with a penalty rate of 1.5% and 1.75%.	• Canada advocated for a "window" in the GEF, which allows for targeted funding for specific issues, such as adaptation, capacity-building, or technology transfer in least developed countries (LDCs), and Small Island Developing States (SIDs). • Canada supported phase-in of forest management under Kyoto Protocol as carbon sequestration activities. • Canada pointed out that the scale of land use, land-use change and forestry (LULUCF) projects are limited in terms of methodologies and implementation. • Canada supported emission trading and flexible mechanisms to meet reduction goals.	• Proposed phase-in forestry management under Kyoto Protocol Article 3.4. • Proposed a special "window" in the GEF to support climate change initiatives. • EU push for legally binding emission reduction commitments, while US push for carbon sinks measures. • No agreement was reached, and negotiations were pushed for COP6.5 (Bonn, Germany, 2001).

COP Number	City, Country	Attendance (Parties + Observer States + Observer Organizations)	Size of the Canadian Delegation (Number of Participants/ Total of Party Participants)	Highlights of the COP	Canada's Contribution	Outputs/Outcomes of the COP
COP 6.5 (2001)	Bonn, Germany	N/A	N/A	• Discussion on forests, croplands, and grazing lands could be used to offset emissions under the Kyoto Protocol. • Discussion on whether nuclear energy projects should count toward Kyoto targets, developed countries "refrain" from using emission reduction units (ERUs)/CERs generated from nuclear facilities to meet Article 3.1. • Russia successfully negotiated higher allowances for its forest-based carbon credits. • Developed countries pledged $410 million per year by 2005 to support developing countries through Kyoto's climate funds. • Senegal played a leading role in advocating for predictable and long-term financial commitments. • Developed countries emphasized cost-effective emissions reduction strategies, particularly through emissions trading and CDM investments.	• Canada, Australia, and Japan secured recognition of forest management credits, while the EU pushed for stricter limits. • Canada advocated for flexible approach and market mechanisms to meet Protocol commitment, while the EU insisted on limits to ensure domestic emissions reductions. • Canada pushed for forest management to be recognized as an eligible activity under Kyoto's carbon accounting rules.	• Bonn agreement was adopted, providing the finalized rules for implementing the Kyoto Protocol, clearing the way for ratification. • Developed countries pledged US$410 million per year to support developing countries. • The European Union pushed for restrictions on carbon trading, while Canada and Japan pushed for more flexibility. • Nuclear projects were excluded from CDM and JI, despite support from Canada, Japan, and Russia. • Early Warning System for Non-Compliance was discussed to notify countries at risk of failing their Kyoto targets before penalties were enforced.

COP Number	City, Country	Attendance (Parties + Observer States + Observer Organizations)	Size of the Canadian Delegation (Number of Participants/ Total of Party Participants)	Highlights of the COP	Canada's Contribution	Outputs/Outcomes of the COP
COP7 (2001)	Marrakech, Morocco	179 states, comprising 1,813 participants 2 observer states, comprising 6 participants 254 observer organizations, comprising 1,723 participants 332 media, comprising 1,086 participants	75	• Discussed the rules for implementing the Kyoto Protocol, particularly in emissions trading, the compliance mechanism, and financial and technical support for developing countries. • Addressed the need for additional funding mechanisms for developing countries beyond the GEF, including the Special Climate Change Fund (SCCF) and the Adaptation Fund. • Reviewed flexible mechanisms in carbon trading, such as CDM and JI projects, allowing developed countries to fund emissions-reduction projects in developing countries. Emphasis was placed on technology transfer to developing countries. • Examined the role of the Consultative Group of Experts (CGE) in improving reporting requirements for developing countries (non-Annex I Parties). Discussions also covered the composition of the expert group. • Debated the inclusion of scientific literature in languages other than English in climate discussions, particularly in the IPCC's Third Assessment Report (TAR). • Indigenous Peoples' Organizations at COP7 requested special status in the climate change negotiation process and proposed the creation of an ad hoc intersessional open-ended working group to enhance Indigenous participation.	• Canada pushed for binding compliance rules under COP/MOP 1, particularly concerning the implementation, procedures, and rules for compliance with the Kyoto Protocol. • Canada proposed a regional workshop to be held in Calgary on integrated climate change impacts and adaptation.	• Finalized the Kyoto Protocol's implementation framework, including Russia's successful negotiation to increase its forest management sinks allowance to 33 megatons of carbon per year. • Set clearer emissions reduction rules, defining how countries could count carbon sinks and emissions trading toward their Kyoto targets. • Established additional funding mechanisms, such as the SCCF and the Adaptation Fund, to provide better support to developing countries. • Introduced penalties for countries failing to meet their Kyoto targets, in which 1.3 times of the excess emissions would be deducted from a country's future carbon budget, following the principle of common but differentiated responsibilities.

COP Number	City, Country	Attendance (Parties + Observer States + Observer Organizations)	Size of the Canadian Delegation (Number of Participants/ Total of Party Participants)	Highlights of the COP	Canada's Contribution	Outputs/Outcomes of the COP
				• A decision was adopted to improve women's participation in UNFCCC bodies. Parties were encouraged to nominate women for elective positions in climate-related decision-making bodies. • Linked climate change to public health impacts, including a request for the World Health Organization (WHO) to provide more information on climate-related diseases.		
COP8 (2002)	New Delhi, India	167 states, comprising 1,456 participants 3 observer states, comprising 12 participants 213 observer organizations, comprising 2,089 participants 222 media, comprising 795 participants	55/1,456	• Delhi reaffirmed the connection between climate change and poverty, calling for sustainable development strategies to incorporate adaptation and mitigation efforts. • The subsidiaries discussed data confidentiality, technical standards for data exchange, and future registry design under Article 7.4. • The European Union proposed an accounting system using temporary certified emission reduction units (TCERs) to address non-permanence in land-use and forestry projects, while G77/China suggested a project-by-project approach.	• Canada reaffirmed its support for the rules and procedures established under the Marrakesh Accords. • Canada supported the scientific community, including the IPCC, and was invited to share information on peer-reviewed research related to emission reduction targets. • Canada and Senegal jointly proposed the "New Delhi Work Program," a five-year work program under Article 6 focused on climate change education, training, and public awareness.	• Guidelines for developing countries in national communications were adopted to improve transparency and reporting. • Article 7.4 (communication) provisions were adopted, but some sections from Articles 7 and 8 remained unresolved. • The SCCF and the Least Developed Countries Fund (LDCF) were officially operationalized. • Parties agreed to peer review and collaboration with the IPCC on future work on climate change. • A voluntary fund for GCOS was proposed, but no formal financial mechanism was established.

COP Number	City, Country	Attendance (Parties + Observer States + Observer Organizations)	Size of the Canadian Delegation (Number of Participants/ Total of Party Participants)	Highlights of the COP	Canada's Contribution	Outputs/Outcomes of the COP
				• Discussion on leakage from activity displacement was a concern, particularly in relation to emission reductions in one area, leading to increases elsewhere. • Brazil proposed limiting crediting periods to 20 years, arguing for historical emissions-based differentiated targets, but Canada, the United States, and Australia raised concerns about a lack of rigorous research on its effectiveness. • Croatia, supported by Yugoslavia, introduced a technical paper on setting a new base year under Article 4.6, but SBSTA concluded it was not consistent with IPCC "good practice." • The Global Climate Observation System (GCOS) discussions focused on climate impact, vulnerability, and adaptation, with Australia supporting a voluntary donor fund. • A proposed web-based system for self-evaluation of climate policies was discussed, but developing countries expressed concerns about the lack of information on adverse effects. • Discussions on hydrofluorocarbons (HFCs) and perfluorocarbons (PFCs) were scheduled to continue until COP11.	• Canada supported holding regional workshops on adaptation and vulnerability assessment to help developing countries.	• The proposed Joint Liaison Group to improve cooperation between UNFCCC, UNCCD, and CBD was supported but not fully operationalized. • No agreement was reached on long-term crediting for afforestation and reforestation under the CDM.

COP Number	City, Country	Attendance (Parties + Observer States + Observer Organizations)	Size of the Canadian Delegation (Number of Participants/ Total of Party Participants)	Highlights of the COP	Canada's Contribution	Outputs/Outcomes of the COP
COP9 (2003)	Milan, Italy	167 states, comprising 1,931 participants 4 observer states, comprising 16 participants 312 observer organizations, comprising 2,698 participants 190 media, comprising 506 participants	66/1,931	• Discussion on utilization of Adaptation Fund outlined in COP7 support developing countries in adapting to climate change. The fund was designated for capacity-building through technology transfer. • Adopted guidelines on afforestation and reforestation activities under the CDM, providing guidelines for LULUCF projects. • Discussions led to further guidance for the operation of the SCCF, focusing on supporting projects related to adaptation, technology transfer, and capacity-building in developing countries. • Discussions on creating a better system for sharing climate data between countries focused on technical standards that would ensure compatibility between different reporting systems. • Adopted the IPCC's Good Practice Guidance (GPG) for LULUCF, guidance on improving the accuracy and transparency of GHG inventories related to forests and land use. • Proposed small-scale afforestation and reforestation projects eligible under the CDM; this was intended to help community-driven and rural projects access carbon markets and financial support.	• Canada pushed for rules that recognized the continued storage of carbon in wood products. • Canada supported the exports of cleaner energy under Protocol Article 2.3. • Canada worked with other countries to draft standardized ways to report on emissions from land use and forestry.	• Finalized Rules for Forestry Projects in the CDM. The system introduced TCERs and long-term certified emission reductions (LCERs) to address concerns about the non-permanence of forest carbon storage. • Approval of the IPCC's GPG for LULUCF. New guidelines were adopted to improve how countries measure and report emissions from forests, agriculture, and land use. • Guidelines were finalized for the SCCF and the LDCF was operationalized, with a focus on helping developing countries adapt to climate change and access private sector in advancing cleaner technologies. • Rules were introduced to make it easier for small-scale afforestation and reforestation projects to participate in carbon trading. • No agreement was reached to submit updates on a fixed schedule to National Climate Reports, with developing countries preferring flexibility.

COP Number	City, Country	Attendance (Parties + Observer States + Observer Organizations)	Size of the Canadian Delegation (Number of Participants/ Total of Party Participants)	Highlights of the COP	Canada's Contribution	Outputs/Outcomes of the COP
				• Discussion on climate change and forestry projects could contribute to the spread of invasive alien species (IAS), and genetically modified trees and crops in afforestation and reforestation projects under the CDM. • Discussed how to account for carbon stored in harvested wood products (HWP), such as lumber and furniture.		
COP10 (2004)	Buenos Aires, Argentina	167 states, comprising 2,210 participants 2 observer states, comprising 9 participants 272 observer organizations, comprising 3,147 participants 240 media, comprising 785 participants	70/2,210	• 10th anniversary of COP: Review of progress achieved over the past decade. • Discussions about transfer of technology and capacity building. • The United States said that individual extreme weather events cannot be linked to climate change. • China raised concerns about the transparency and reliability of the International Transaction Log (ITL). • Concerns raised over GEF co-financing rules made it harder for vulnerable countries to access funds. • Discussion about LDCF, but no decision made. • Russia officially ratified the Kyoto Protocol.	• Canada backed Australia's proposal for open-ended policy dialogue to distinguish between human-induced and natural effects in LULUCF activities. • The agenda relating to cleaner or less-greenhouse-gas emitting energy was withdrawn after Canada retracted its proposal on cleaner energy exports. • Canada pushed to remove text on LDCF guidance, arguing more time was needed to refine the operational framework.	• Administrator of ITL was requested to conduct thorough testing and an independent assessment. • GEF urged to become more flexible, simplify access to funds, and provide support for small-scale projects in SIDS. • The Buenos Aires Programme of Work on Adaptation and Response Measures laid the groundwork for a future five-year adaptation plan. • The Buenos Aires Programme of Work on Adaptation and Response Measures launched to strengthen adaptation, mitigation, technology transfer, and capacity-building efforts. • The Russian Federation ratified the Kyoto Protocol.

COP Number	City, Country	Attendance (Parties + Observer States + Observer Organizations)	Size of the Canadian Delegation (Number of Participants/ Total of Party Participants)	Highlights of the COP	Canada's Contribution	Outputs/Outcomes of the COP
COP11 (2005)	Montreal, Canada	181 states, comprising 2,804 participants 2 observer states, comprising 5 participants 420 observer organizations, comprising 5,848 participants 287 media, comprising 817 participants	321/2,804	• First conference of countries committed to the Kyoto Protocol to address its implementation and operational details. • Discussion on improving the efficiency and credibility of the CDM. • Belarus sought an emissions target under the Kyoto Protocol, but no agreement was reached. • Debate about the adoption of a five-year program of work, focusing on adaptation. • Discussion about the procedure for appointing an executive secretary. • Continued discussion from COP10 on accounting rules. • The United States walked out against discussions on future commitments but later re-engaged in non-binding commitments.	• Conference President Dion urged delegates to consider long-term cooperative action beyond 2012. • Canada emphasized a science-based approach and long-term cooperation in shaping the five-year program of work on adaptation. • Canada supported a sectoral approach to mitigation, prioritizing transport, renewable energy, energy efficiently, and carbon management. • Canada pledged US$1.5 million to the CDM, the largest contribution at COP11. • Canada cautioned that Saudi Arabia's proposal to amend the compliance system could create uncertainty and possibly divide parties into two categories. • Canada pledged CAN$5 million for the Adaptation Fund. • Canada proposed recognizing HFC-23 destruction projects to count toward emission reduction targets, with strict conditions.	• The Marrakesh Accords were formally adopted, finalizing the operational rules for the Kyoto Protocol. • Parties pledged US$8.29 million to fund the CDM Executive Board's financing shortfall. • The Five-Year Programme of Work on Adaptation was adopted, focusing on resilience and climate risk management. • LULUCF accounting rules were refined, excluding certain carbon stock calculations. • The JI Supervisory Committee was created to oversee JI projects under the Kyoto Protocol. • Parties agreed that excess emissions in one period would result in a 1.3x education in the next commitment period's assigned amount. • Established the Ad Hoc Working Group on Further Commitments (AWG-KP) to advanced discussions on post-2012 climate action.

COP Number	City, Country	Attendance (Parties + Observer States + Observer Organizations)	Size of the Canadian Delegation (Number of Participants/ Total of Party Participants)	Highlights of the COP	Canada's Contribution	Outputs/Outcomes of the COP
					• Canada supported a "two-track" strategy for post-2012 action, maintaining parallel tracks under both the UNFCCC and the Kyoto Protocol.	
COP12 (2006)	Nairobi, Kenya	180 states, comprising 2,844 participants 3 observer states, comprising 8 participants 298 observer organizations, comprising 2,933 participants 282 media, comprising 663 participants	55/2,844	• Discussion about whether mitigation or adaptation should be assigned higher priority and a greater share of financing. • Debate over whether to extend the Expert Group on Technology Transfer (EGTT) or replace it with a new mandate. • Discussion on moving forward with the Buenos Aires Programme of Work, but no formal implementation was agreed upon. • Discussion on including carbon capture and storage (CSS) under CDM. • Discussion about adoption of rules of procedure for Joint Implementation Supervisory Committee (JISC). • Discussion about issuing credits for CERs, the destruction of HFC-23; no agreement was reached. • Discussions on possible changes to Kyoto Protocol mechanisms, including funding and compliance rules.	• Canada emphasized limiting the discussion to initial activities and opposed China's proposal to establish an advisory group for the Five-Year Programme of Work on Adaptation. • Canada expressed interest in accessing CCS technology. • Canada suggested a country-driven approach, efficiency and effectiveness, and knowledge and networking capacity as the Adaptation Fund's principles.	• Belarus was allowed to take on voluntary emission reduction commitments under the Kyoto Protocol. • Adaptation was given priority over mitigation, receiving greater focus on funding and initiatives. • Changes made to the GEF to make climate financing more accessible to developing countries. • With no agreement on its future, EGTT was extended for another year, leaving many parties disappointed. • No agreement on formal implementation of Bruno's Aires Program of Work on Adaptation and Response Measures. • No consensus was reached on CCS methodologies, leaving its role in CDM unresolved. • Rules of procedure and project design documents adopted for JISC.

COP Number	City, Country	Attendance (Parties + Observer States + Observer Organizations)	Size of the Canadian Delegation (Number of Participants/ Total of Party Participants)	Highlights of the COP	Canada's Contribution	Outputs/Outcomes of the COP
COP13 (2007)	Bali, Indonesia	188 states, comprising 3,508 participants 4 observer states, comprising 8 participants 413 observer organizations, comprising 5,815 participants 531 media, comprising 1,498 participants	60/3,508	• Discussion on new mandate for Consultative Group of Experts (CGE). • Discussions about issuing CERs for the destruction of HFC-23 under CDM. • Discussion about the institutional arrangements for the Adaptation Fund. • Discussion about the Bali Roadmap, a two-year plan to shape the post-2012 climate framework. • Debate over Russia's proposal to develop procedures for voluntary commitments, though it was not adopted.	• Canada expressed disappointment on the outcome of no agreement on the mandate for the CGE. • Canada rejected the inclusion of the 25%–40% emission reduction target in the AWG's report, arguing for broader commitments from major emitters. • Canada suggested that parts of the Russian proposal could be considered as part of the Bali Roadmap.	• The Fourth Assessment Report (AR4) of the IPCC was recognized as the most authoritative assessment of climate change. • The GEF was adjusted to provide full-cost funding for certain adaptation projects and simplify its procedures for developing countries. • COP adopted the amended New Delhi work program and extended it for five years. • Revised limit for afforestation and reforestation project activities from 8 kt to 16 kt of CO_2 per year. • Establishment of Ad Hoc Working Group on Long-Term Cooperative Action to enable long-term cooperation on mitigation beyond 2012. • The IPCC AR4 reported that global emissions should peak within 10 to 15 years and decline 25%–40% below 1990 levels by 2020, but this was not formally adopted. • Produced the Bali Roadmap, a two-year framework for all countries to respond to climate challenges. • Operationalized the Adaptation Fund to receive funding through CDM projects.

COP Number	City, Country	Attendance (Parties + Observer States + Observer Organizations)	Size of the Canadian Delegation (Number of Participants/ Total of Party Participants)	Highlights of the COP	Canada's Contribution	Outputs/Outcomes of the COP
COP14 (2008)	Poznań, Poland	189 states, comprising 3,958 participants 2 observer states, comprising 9 participants 464 observer organizations, comprising 4,466 participants 371 media, comprising 819 participants	46/3,958	• Review of the implementation of commitments relating to the financial mechanism, technology transfer, capacity building, and adverse effects and response measures. • Discussion about renewing the mandate of the CGE. • Concern from developing countries about GEF multiple sources of funding outside the convention. • Discussion about including Indigenous Peoples and recognizing Indigenous Peoples in climate agreements. • Discussion about helping countries with setting targets of emission reduction. • Discussion about expanding climate financing by applying a 2% levy on carbon trading to fund climate adaptation.	• Canada rejected the creation of new financial systems and wanted the improvement of existing institutions instead. • Canada supported maximizing existing institutions inside and outside of the Convention to attract investments from parties to support the development and transfer of mitigation and adaptation technologies.	• Poznan Strategic Programme on Technology was a substantive decision. The primary focus was on scaling-up investment in technology transfer through the existing GEF. • GEF was asked to improve how it works with its partner organizations to help least developed countries get funding more quickly. This includes setting deadlines. • COP14 approved a decision that encouraged countries to start full negotiations in 2009 to create a new climate agreement. • Countries could not agree on renewing the CGE and on creating new financial systems • Countries could not agree on expanding financing by applying 2% levy on carbon trading to fund climate adaptation.

COP Number	City, Country	Attendance (Parties + Observer States + Observer Organizations)	Size of the Canadian Delegation (Number of Participants/ Total of Party Participants)	Highlights of the COP	Canada's Contribution	Outputs/Outcomes of the COP
COP15 (2009)	Copenhagen, Denmark	194 states, comprising 10,583 participants 2 observer states, comprising 8 participants 900 observer organizations, comprising 13,482 participants 1,287 media, comprising 3,221 participants	204/10,583	• Discussions on adopting a "COP accord" drafted by the representative group of leaders. • Discussion about limiting global temperature rise to below 2°C, with a possible review for a 1.5°C target. • Discussion about creating a legally binding treaty that provides rules for LULUCF. • Discussion about developed countries providing immediate climate aid to developing countries. • Suggestion by developed countries for all major economies to adopt legally binding emission reduction targets. • Suggestion by the European Union and island nations to adopt a legally binding treaty to replace the Kyoto Protocol.	• Canada opposed the amendment of Article 3.9 of the Kyoto Protocol (included mandatory emission targets for developed countries). • Canada preferred non-binding language like "encouraging parties" to promote CDM projects in underrepresented regions. • Canada supported the implementation of the Copenhagen Accord.	• The Copenhagen Accord was to be implemented voluntarily. • Voluntary emission reduction pledges were agreed on. • Developed countries agreed to provide developing countries with $30 billion in immediate funding (2010–2012) and set a long-term goal of $100 billion per year by 2020. • There were no clear rules on how the $100 billion would be distributed.

COP Number	City, Country	Attendance (Parties + Observer States + Observer Organizations)	Size of the Canadian Delegation (Number of Participants/ Total of Party Participants)	Highlights of the COP	Canada's Contribution	Outputs/Outcomes of the COP
COP16 (2010)	Cancún, Mexico	192 states, comprising 5,183 participants 2 observer states, comprising 9 participants 687 observer organizations, comprising 5,386 participants 568 media, comprising 1,270 participants	88/5,183	• The focus in Cancún was on a two-track negotiating process, aiming to enhance long-term cooperation under the Convention. • Discussion about why COP15 was a failure. • Discussion about the consideration and inclusion of Indigenous Peoples in climate policy and mitigation. • Discussion about the nature of mitigation and whether it should involve targets or commitments. • Developing countries requested that Annex-I countries contribute 6% of their GNP for climate change mitigation and adaptation and 1% for forest-related activities as repayment for their "climate debt." • A proposal to expand priority recipients beyond LDCs, SIDS, and Africa to include regions prone to droughts, floods, desertification, and extreme climate events. • Developing countries preferred a 1.5% GDP contribution from developed countries over the $100 billion annual target by 2020.	• Canada supported addressing the legal status of Compliance Committee members through the SBI's work on privileges and immunities. • Canada acknowledged the importance of a comprehensive approach to Croatia's appeal regarding its Kyoto Protocol base-year emissions.	• Developed countries agreed to submit emission reduction targets. • Developing countries committed to nationally appropriate mitigation actions (NAMAs), with reporting and verification procedures for both. • Creation of the Cancún Adaptation Framework and an Adaptation Committee to help vulnerable countries adapt to climate change. • Establishment of a Technology Mechanism, including the Technology Executive Committee (TEC) and the Climate Technology Centre and Network (CTCN) to support the development and transfer of clean technologies. • COP16 restored trust in the multilateral process after the Copenhagen Accord's disappointment in 2009. It laid a solid foundation for future negotiations, especially regarding finance, adaptation, and technology. • Did not end with a legally binding deal but uncertainty around Kyoto's future. Left critical issues unresolved until later conferences like COP17 in Durban (2011).

COP Number	City, Country	Attendance (Parties + Observer States + Observer Organizations)	Size of the Canadian Delegation (Number of Participants/ Total of Party Participants)	Highlights of the COP	Canada's Contribution	Outputs/Outcomes of the COP
COP17 (2011)	Durban, South Africa	192 states, comprising 5,399 participants 3 observer states, comprising 14 participants 761 observer organizations, comprising 5,811 participants 507 media, comprising 1,265 participants	69/5,399	• Discussions about managing risks of extreme weather events and disasters to advance climate adaptation. • Discussion around taking into account challenges related to gender and disability in mitigation and capacity building. • Discussion on the integration of CCS and the CDM under the Kyoto Protocol. • Discussion on emissions reductions on fuel consumption in aviation and marine transportation. • The European Union suggested a new global binding framework on matters to funding developing countries, and reaffirmed its commitment to the mobilization reaffirmed the previous commitment made by the international community to mobilise US$100 billion annually by 2020.	• Canada supported a proposal by New Zealand to include a template for understanding underlying assumptions for NAMAs. • Canada co-chaired an informal subcommittee to address the report of the compliance committee. • Canada co-chaired a committee that discussed how safeguards are addressed and respected, modalities for forest reference emission levels, and forest reference levels.	• Countries agreed to negotiate a new global climate agreement by 2015, with implementation starting in 2020. This marked a significant step toward legally binding commitments for all states. • COP17 established a framework for a new carbon market mechanism beyond the CDM, allowing for international trading of emission reduction credits. • A decision was made to include CCS projects under the CDM, allowing countries to earn carbon credits from storing CO_2 underground. • The issue of compensating developing countries for climate-related loss and damage was formally acknowledged, though no binding financial commitments were made. • Countries recognized the need to scale up their commitments to keep global warming below 2°C, though no binding emissions targets were set.

COP Number	City, Country	Attendance (Parties + Observer States + Observer Organizations)	Size of the Canadian Delegation (Number of Participants/ Total of Party Participants)	Highlights of the COP	Canada's Contribution	Outputs/Outcomes of the COP
COP18 (2012)	Doha, Qatar	189 states, comprising 4,343 participants 3 observer states, comprising 13 participants 631 observer organizations, comprising 3,965 participants 344 media comprising 683 participants	53/4,343	• The first UN climate negotiations to be held in the Middle East. • Focused is placed on implementing the outcomes of the Kyoto Protocol. • Negotiations surrounding climate finance, how to scale up funding, and make funds more accessible to developing countries. • Marked the end of the first commitment period under the Kyoto Protocol, and launched the second commitment period. • Further development of JI and CDM procedures. • Controversy over the selling of assigned amount units (AAUs). • The issue of gender balance in climate negotiations was considered by the Subsidiary Body of Implementation.	• Canada opposed an initial informal overview text of the AWG-LCA, asking for greater recognition of the progress made by the AWG-LCA in establishing new institutions. • Canada co-chaired discussions regarding the controversial issue of loss and damage through the Subsidiary Body of Implementation. • Canada co-chaired discussions regarding methodological guidance on REDD+, advocating for the need to find a solution in Doha. • Canada co-chaired discussions regarding agriculture through the Subsidiary Body for Scientific and Technical Advice.	• Adoption of the Doha Climate Gateway package, which receives mixed reviews from delegates. • An amendment was made to the Kyoto Protocol that includes new emission reduction commitments for the second period. • Nitrogen trifluoride is added to the Kyoto Protocol as a GHG. • Upon recommendation from the SBI, COP adds the issue of gender and climate to the standing agenda. • The work of the AWG-LCA is deemed complete, and it is terminated.

COP Number	City, Country	Attendance (Parties + Observer States + Observer Organizations)	Size of the Canadian Delegation (Number of Participants/ Total of Party Participants)	Highlights of the COP	Canada's Contribution	Outputs/Outcomes of the COP
COP19 (2013)	Warsaw, Poland	190 states, comprising 4,011 participants 2 observer states, comprising 11 participants 681 observer organizations, comprising 3,695 participants 339 media, comprising 658 participants	56/4,011	• The issue of adaptation dominated the conference, especially regarding the implementation of the Ad Hoc Working Group on the Durban Platform for Enhanced Action (ADP). • Discussions regarding long-term finance and how to get more funding for adaptation in developing countries. • Concerns are voiced regarding the viability of current JI institutions and the sustainability of the Adaptation Fund (due to a lack of contributions from developed countries). • Workshop on Urbanization and Climate Change in Cities. • Tensions and disagreements arose surrounding the ADP, with delegates butting heads over the role of the ADP as supplementary to the existing climate regime, not as a new regime itself. • Delegates from the Philippines protested the slow-moving climate deliberations at COP19.	• Canada highlighted the importance of intellectual property rights in the innovation of climate technology, but opposed their inclusion in the ADP. • Canada noted that finances for adaptation must be mobilized from the private sector as well, as public funds alone cannot provide for the countries most in need. • Canada supported the US call for mitigation efforts to come from the sub-national level as well as the national level, and the need for collaboration between the two levels. • Under the SBSTA, Canada co-facilitated talks regarding the Nairobi Work Program (a knowledge-to-action initiative). • Canada co-chaired talks regarding procedures and monitoring activities for REDD+. • Canada, along with other developed countries, opposed the Brazilian proposal to consider historical responsibilities under the SBSTA.	• Resolutions are made to speed up mitigation efforts and set higher targets leading up to 2020. Discussions regarding decision-making mechanisms within the UNFCCC are fruitless, and the matter is ultimately pushed to COP20. • The Warsaw International Mechanism is developed under the Cancun Adaptation Framework, dealing specifically with loss and damage related to climate in developing countries. • Multiple decisions are made regarding REDD+, and are known as the Warsaw REDD+ Framework.

COP Number	City, Country	Attendance (Parties + Observer States + Observer Organizations)	Size of the Canadian Delegation (Number of Participants/ Total of Party Participants)	Highlights of the COP	Canada's Contribution	Outputs/Outcomes of the COP
					• Canada confirmed its commitment to the joint mobilization of USD\$100 billion annually to the GCF by 2020 but noted that these funds may come from more than just the public purse.	
COP20 (2014)	Lima, Peru	190 states, comprising 6,809 participants 2 observer states, comprising 8 participants 772 observer organizations, comprising 4,654 participants 472 media, comprising 1,060 participants	67/6,809	• Intended nationally determined contributions (INDCs) were a hot topic of discussion. • A focus was placed on the rebuilding of transparency, confidence, and trust between countries in the lead-up to COP21. • Significant work was undertaken to advance the Durban Platform for Enhanced Action, resulting in the Lima Call for Climate Action. • Activities under JI were reported to be in decline, and the importance of JI for mitigation efforts was reaffirmed. • Increased use of the CDM as a mitigation tool was highly encouraged. • The information hub committed to in Warsaw was officially named the Lima Information Hub for REDD.	• Canada was heavily involved in the negotiations to refine the text of the Lima Call for Climate Action, favouring the use of more ambiguous language and a looser framework. • Canada co-led talks regarding the Warsaw International Mechanism under the SBI/ SBSTA, leading to a multitude of procedural decisions being achieved. • Canada co-led talks regarding the Nairobi Work Program under the SBSTA. • Canada was thanked by the SBI for providing financial support to the Least Developed Countries Expert Group (LEG). • Canada co-chaired a contact group on long-term climate finance and the SCF.	• Foundation is laid for a new agreement to be reached in Paris at COP21. • The Lima Call for Climate Action is established. • The Lima Work Program on Gender, a two-year program focused on promoting gender balance in the UNFCCC and creating gender-sensitive climate policy, was established. • Discussions regarding linking the technology mechanism and the financial mechanism of the convention are fruitless, as delegates cannot reach an agreement; the issue is pushed to COP21.

COP Number	City, Country	Attendance (Parties + Observer States + Observer Organizations)	Size of the Canadian Delegation (Number of Participants/ Total of Party Participants)	Highlights of the COP	Canada's Contribution	Outputs/Outcomes of the COP
COP21 (2015)	Paris, France	196 states, comprising 19,208 participants 2 observer states, comprising 52 participants 1,203 observer organizations, comprising 8,314 participants 1,176 media, comprising 2,798 participants	287/19,208	• The bulk of negotiations revolved around the creation of the Paris Agreement, a package of decisions made under the ADP that would determine international climate policy for the following decade. • INDCs remained an issue of contention between developed and developing countries. • The role of local communities and Indigenous Peoples was highlighted. • Differentiation was highlighted as an important aspect to consider in the new agreement. • Disagreement between parties on whether or not to frame climate change as an isolated issue, or as part of a complex network relating to human rights. • The issue of transparency was given attention in drafting the agreement.	• Canada co-led discussions on cooperative approaches and mechanisms under the Comité de Paris.	• The Paris Agreement is established, creating an international treaty that reaffirms parties' dedication to combatting the impacts of climate change. • The Paris Agreement Implementation and Compliance Committee (PAICC) and the Paris Committee on Capacity-Building (PCCB) are established as instruments of the Paris Agreement.

COP Number	City, Country	Attendance (Parties + Observer States + Observer Organizations)	Size of the Canadian Delegation (Number of Participants/ Total of Party Participants)	Highlights of the COP	Canada's Contribution	Outputs/Outcomes of the COP
COP22 (2016)	Marrakech, Morocco	194 states, comprising 15,978 participants 1 observer state, comprising 7 participants 992 observer organizations, comprising 5,475 participants 572 media, comprising 1,204 participants	202/15,978	• Paris Agreement Implementation, countries agreed to finalize rulebook by 2018. • Marrakech Action Proclamation, a statement issued to further strengthen commitments to climate action. • Decision to make the Adaptation Fund serve the Paris Agreement. • Approval of five-year work plan for the Warsaw International Mechanism (WIM) to address climate-induced damage and loss. • Continuation of Lima Work Program, with focus on gender-responsive climate action. • Launch of PCCB. • Developed countries presented $100 billion to support developing countries.	• Canada reaffirmed its commitment to contribute $2.65 billion by 2021 to help developing countries to transition to low-carbon economies. • Canada actively supported the development of the Lima Work Program. • Canada highlighted the importance of Indigenous Knowledge and practices in climate change. • Canada supported several initiatives targeting climate resilience in vulnerable African nations.	• Clear framework for operating the Paris Agreement. • Reaffirmation of global solidarity and momentum for climate action. • Better funding structure for developing countries' adaptation. • Clear structures to address climate-related damages and losses. • Strengthened gender-responsive climate action. • Capacity building to support developing countries' capabilities. • Greater transparency and clarity on finances.

COP Number	City, Country	Attendance (Parties + Observer States + Observer Organizations)	Size of the Canadian Delegation (Number of Participants/ Total of Party Participants)	Highlights of the COP	Canada's Contribution	Outputs/Outcomes of the COP
COP23 (2017)	Bonn, Germany (presided over by Fiji)	194 states, comprising 9,196 participants 1 observer state, comprising 6 participants 1,084 observer organizations, comprising 5,543 participants 539 media, comprising 1,283 participants	148/9,196	• Talanoa dialogue launched a process to assess progress on climate goals. • Local Communities and Indigenous Peoples Platform created to aid engagement of Indigenous Knowledge. • First COP to be presided over by an island state. • Gender Action Plan created to integrate gender equality into climate policy solutions. • Ocean Pathway Initiative launched to address ocean health within climate change. • The Suva expert dialogue was launched to support loss and damage knowledge.	• Canada announced funding support for Small Island Developing States (SIDS). • Canada showcased its support for the adoption of the Gender Action Plan. • Canada advocated for the integration of Indigenous Knowledge into climate policy.	• Improved structure for transparent dialogue. • Further inclusive decision-making and climate solutions. • Improved representation and inclusion of women in climate negotiations. • Greater recognition of oceans' role in climate regulation. • Increased focus on practical solutions for loss and damage due to climate.
COP24 (2018)	Katowice, Poland	196 states, comprising 11,090 participants 1 observer state, comprising 10 participants 1,148 observer organizations, comprising 6,193 participants 566 media, comprising 1,126 participants	138/11,090	• Agreement and implementation of the Paris Agreement Work Program (PAWP) • Established a set of rules for reporting NDCs. • Operationalized and facilitated a working group to recognize Indigenous Knowledge. • Global stock-take defined, assessed every five years.	• Canada endorsed the Just Transition Silesia Declaration. • Canada supported the operationalization of Local Communities and Indigenous Peoples Platform.	• Clear operational guidelines for implementing the Paris Agreement. • Established the Just Transition Declaration. • Enhanced transparency and coordination of national commitments. • Stronger social protections for workers. • Enhanced recognition of Indigenous Knowledge. • Structured approach to measure global progress.

COP Number	City, Country	Attendance (Parties + Observer States + Observer Organizations)	Size of the Canadian Delegation (Number of Participants/ Total of Party Participants)	Highlights of the COP	Canada's Contribution	Outputs/Outcomes of the COP
COP25 (2019)	Madrid, Spain (presided over by Chile)	196 states, comprising 11,406 participants 1 observer state, comprising 8 participants 1,176 observer organizations, comprising 8,775 participants 844 media, comprising 2,165 participants	157/11,406	• Decision to push stronger national climate commitments by 2020. • Operationalized the Facilitative Working Group with a two-year work plan. • Continued promotion of non-state actor involvement. • Agreement to hold round tables on pre-2020 COPs. • Continued focus on responsibilities and unfulfilled commitments.	• Canada showed support for technical assistance to help vulnerable states. • Canada supported and enhanced Gender Action Plan. • Canada contributed to negotiations on the Paris Agreement, emphasizing the need for clearer reporting mechanisms. • Canada backed the decision to further discussions on ocean climate change.	• Maintained momentum for Paris Agreement's implementation. • Enhanced support for vulnerable countries. • Established the Santiago Network. • Strengthened Indigenous participation in climate discussions. • Extended the Marrakech Partnership to 2025 with new mandates.
COP26 (2021)	Glasgow, Scotland (presided over the United Kingdom)	194 states, comprising 9,742 participants 1 observer state, comprising 7 participants 1,725 observer organizations, comprising 11,000 participants 1,089 media, comprising 2,602 participants, 15,106 additional badges (party overflow, Global Climate Action and Momentum for Change, staff, other).	211/9,742	• First-ever UNFCCC reference to phasing down unabated coal power and phasing out inefficient fossil fuel subsidies. • Calls for developed countries to double adaptation finance from 2019 levels by 2025. • Requirement for countries that have not yet updated their NDCs to do so before COP27. • Glasgow Dialogue was launched to discuss loss and damage financing between 2022 and 2024. • Operationalization of carbon credit trading under Article 6. • Agreement on common time frames for NDCs (countries submit new NDCs in 2025, valid until 2035, then in 2030 for 2040, and so on).	• Canada pledged to increase its climate finance contributions, focusing on support for developing countries. • Canada co-led the Global Methane Pledge, committing to reducing methane emissions. • Canada announced strengthened climate policies, including increased ambition for reducing emissions. • Canada joined the coalition to phase out coal but advocated for support in transition strategies for coal-dependent communities.	• Recognized multilateralism and international cooperation as key to tackling climate change. • Emphasized the urgency of adaptation finance and called for more support from developed countries. • While no dedicated funding was created, the Santiago Network was strengthened to provide technical support. • A two-year dialogue on loss and damage finance was established. • The enhanced transparency framework will ensure regular reporting and accountability on emissions reductions.

COP Number	City, Country	Attendance (Parties + Observer States + Observer Organizations)	Size of the Canadian Delegation (Number of Participants/ Total of Party Participants)	Highlights of the COP	Canada's Contribution	Outputs/Outcomes of the COP
				• Enhanced transparency framework was fully completed, allowing parties to submit biennial transparency reports starting in 2024. • Global Methane Pledge: Over 100 countries agreed to cut global methane emissions by 30% by 2030. • Glasgow Leaders' Declaration on Forests and Land Use signed by 120 countries to halt deforestation by 2030. • Financial commitments: new climate finance pledges, including an $800 million increase in adaptation funding, first-ever US contribution to the Adaptation Fund, and additional pledges from Japan, Germany, and Spain. • Coal phase-out language was watered down to "phasing down unabated coal" after a last-minute intervention by India and China. • Loss and damage funding did not include a financial mechanism, which disappointed vulnerable states. • Developed countries failed to meet their $100 billion climate finance commitment, acknowledging they will only reach the target by 2023.	• Canada emphasized Indigenous Peoples' role in climate action, recognizing their rights and traditional knowledge as part of climate solutions.	• New agreements for carbon trading mechanisms under Article 6 of the Paris Agreement. • Developed countries were urged to double adaptation finance by 2025. • Emphasis on "just transition" for workers in fossil fuel-dependent sectors.

COP Number	City, Country	Attendance (Parties + Observer States + Observer Organizations)	Size of the Canadian Delegation (Number of Participants/ Total of Party Participants)	Highlights of the COP	Canada's Contribution	Outputs/Outcomes of the COP
COP27 (2022)	Sharm el-Sheikh, Egypt	195 states, comprising 11,969 participants 0 observer states 1,809 observer organizations, comprising 12,240 participants 948 media, comprising 2,160 participants, 23,334 additional badges (party overflow, Global Climate Action and Momentum for Change, staff, other).	260/11,969	• A historic decision was made to to establish a Loss and Damage Fund to assist developing countries in responding to the adverse effects of climate change. • Details of the fund's operation will be worked out over the next year. • Reinforced commitments from the Glasgow Climate Pact (COP26) on phasing down unabated coal power and phasing out inefficient fossil fuel subsidies. • Countries that have not yet updated NDCs urged to do so. • A work program on just transition was established to explore pathways for an equitable energy transition. • The Sharm el-Sheikh dialogue was launched to enhance understanding of financial flows for climate-resilient development. • Calls for developed countries to provide enhanced support to assist developing countries in both mitigation and adaptation. • Reforms in multilateral development banks (MDBs) were urged to ensure climate finance is accessible and effectively used.	• Canada supported the call for increased financial support to developing countries, particularly for adaptation. • Canada advocated for reforming MDBs to ensure better climate finance delivery. • Canada reaffirmed commitment to phasing out inefficient fossil fuel subsidies and scaling up clean energy solutions. • Canada supported discussions on phasing down coal while recognizing challenges in transitioning for certain economies. • Canada promoted the inclusion of Indigenous perspectives in climate action and decision-making processes. • Canada engaged in discussions on a work program for a just transition, ensuring workers and communities dependent on fossil fuels have pathways to a sustainable future.	• A transitional committee was established to define the structure and function of the Loss and Damage Fund, set for operationalization at COP28. • Countries will engage in discussions on how to achieve climate goals while ensuring economic and social equity. • Progress was made on rules for carbon credit trading under the Paris Agreement, Article 6. • Calls for reforming international financial institutions and MDBs to support more climate finance effectively. • While there were commitments to continue efforts toward the 1.5°C target, no new agreements were reached on phasing out all fossil fuels, despite push from many countries.

COP Number	City, Country	Attendance (Parties + Observer States + Observer Organizations)	Size of the Canadian Delegation (Number of Participants/ Total of Party Participants)	Highlights of the COP	Canada's Contribution	Outputs/Outcomes of the COP
				• A new mitigation work program was launched to scale up emissions reduction before 2030, seen as the "critical decade" for climate action. • Some countries raised concerns about the effectiveness of these mitigation efforts in keeping the 1.5°C target alive. • Progress on defining a global goal on adaptation (GGA) to help countries strengthen resilience. • Encouraged scaling up financial, technological, and capacity-building support for adaptation.		
COP28 (2023)	Dubai, United Arab Emirates	196 states, comprising 20,204 participants 1 observer state 2,279 observer organizations, comprising 16,747 participants 1,002 media, comprising 2,673 participants, 44,260 additional badges (party overflow, Global Climate Action and Momentum for Change, staff, other).	636/20,204	• The Loss and Damage Fund, established at COP27, was officially operationalized at COP28. • Several countries pledged financial contributions to support vulnerable states. • Some opposition from fossil fuel-producing states regarding the "phase-out" of fossil fuels, with a compromise on "transitioning away." • Recognition of the need for deep, rapid, and sustained reductions in GHG emissions to align with the 1.5°C goal. • States urged to strengthen their NDCs. • Tripling of renewable energy capacity globally and doubling energy efficiency by 2030 was encouraged.	• Canada pledged financial support for the Loss and Damage Fund, aiding vulnerable countries. • Canada supported reforms to international financial systems to make climate finance more accessible. • Canada advocated for phasing down coal and reducing fossil fuel reliance while ensuring energy security. • Canada backed the global tripling of renewable energy capacity and supported investments in clean technologies.	• Loss and Damage Fund moves forward, operationalized with initial funding pledges, though details on long-term financial mechanisms remain undecided. • Continued debate on how to finance long-term loss and damage responses. • Wide support to triple renewable energy and double energy efficiency goals by 2030. • Developed states urged to fulfill their $100 billion annual climate finance pledge, which is still unmet. • Call for new climate finance mechanisms to support developing countries.

COP Number	City, Country	Attendance (Parties + Observer States + Observer Organizations)	Size of the Canadian Delegation (Number of Participants/ Total of Party Participants)	Highlights of the COP	Canada's Contribution	Outputs/Outcomes of the COP

				• Phasing down unabated coal power and transitioning away from fossil fuels in an orderly, just, and equitable manner.
				• Calls for accelerating zero- and low-carbon technologies, including CCS and low-carbon hydrogen.
				• Finalized the GGA framework, which includes:
				  o Impact and vulnerability assessments by 2030.
				  o Multi-hazard early warning systems by 2027.
				  o Climate information services and risk reduction plans.
				• Developed countries failed to meet the $100 billion climate finance goal, which was due by 2020.
				• Calls for reforming MDBs were made to improve financial flows.
				• Highlighted the adaptation finance gap, urging states to double adaptation funding by 2025.
				• Launched a work program on Just Transition Pathways to explore equitable ways to transition to clean energy.
				• Recognition of the need for economic diversification for countries dependent on fossil fuels.
					• Canada emphasized the role of Indigenous Knowledge and leadership in climate adaptation and mitigation strategies.
					• Canada advocated for inclusive and equitable climate policies.
					• Canada reaffirmed commitment to reducing methane emissions, aligning with the Global Methane Pledge.
					• Canada supported stronger international efforts to cut non-CO_2 GHGs.
						• The global stock-take underscored that current actions are insufficient to meet the Paris Agreement goals.
						• States encouraged to enhance their NDCs before COP30 (Brazil, 2025).
						• Countries emphasized the social and economic dimensions of the energy transition.
						• Acknowledgment of the need for financial and technical support for states undergoing fossil fuel transitions.

COP Number	City, Country	Attendance (Parties + Observer States + Observer Organizations)	Size of the Canadian Delegation (Number of Participants/ Total of Party Participants)	Highlights of the COP	Canada's Contribution	Outputs/Outcomes of the COP
COP29 (2024)	Baku, Azerbaijan	193 states, comprising 14,074 participants 0 observer states 1,944 observer organizations, comprising 10,562 participants 940 media, comprising 2,220 participants, 27,292 additional badges (party overflow, Global Climate Action and Momentum for Change, staff, other).	62/14,074	• A key milestone was the agreement on a new collective quantified goal (NCQG) on climate finance, replacing the previous $100 billion per year target. • Developed countries pledged at least $1.3 trillion per year by 2035, with at least $300 billion dedicated to supporting developing countries. • Developed countries are required to lead financial contributions, while developing countries can contribute voluntarily. • Commitment to triple financial outflows from key climate funds (Green Climate Fund, Adaptation Fund, Loss and Damage Fund) by 2030. • Emphasis on public and grant-based finance for adaptation and loss and damage. • After years of negotiation, rules were finalized for the carbon markets under Article 6.2 and 6.4, enabling states to trade carbon credits under strict guidelines. • Measures were introduced to ensure environmental integrity, transparency, and compliance in emissions trading. • No agreement on how to implement the global stock-take outcomes related to energy transition.	• Canada pledged new contributions toward the Loss and Damage Fund. • Canada supported increasing climate finance and backed efforts to reform international financial systems to improve fund accessibility for developing countries. • Canada reaffirmed support for phasing down fossil fuels while ensuring a just transition for workers and communities. • Canada advocated for stronger methane reduction commitments in alignment with the Global Methane Pledge. • Canada highlighted the role of Indigenous Knowledge and leadership in climate action. • Canada called for equity-based climate policies that ensure marginalized groups benefit from climate adaptation initiatives. • Canada pushed for more transparent reporting mechanisms to track climate finance flows and their impact.	• A formal climate finance target was set at $1.3 trillion per year by 2035, with at least $300 billion directed at developing countries. • Key disagreements remained over the distribution and access mechanisms for these funds. • Final rules were agreed upon, allowing countries to trade emissions credits under strict monitoring conditions. • Carbon market participants must ensure real emissions reductions, avoiding double counting and loopholes. • The Loss and Damage Fund was fully operationalized, ensuring financial support for countries most affected by climate change. The governance structure for the Loss and Damage Fund was finalized. The fund will provide financial support to climate-vulnerable countries, including SIDS and LDCs. • New financing mechanisms were introduced to address climate-related disasters. • Developed countries failed to meet the previous $100 billion climate finance goal, raising concerns about future commitments.

COP Number	City, Country	Attendance (Parties + Observer States + Observer Organizations)	Size of the Canadian Delegation (Number of Participants/ Total of Party Participants)	Highlights of the COP	Canada's Contribution	Outputs/Outcomes of the COP
				• Continued debates on the Just Transition Work Program and the effectiveness of the Adaptation Committee.		• Disagreements continued over how to implement the global stock-take recommendations on energy transition and just transition policies. • Countries were urged to update their NDCs before COP30 (Brazil, 2025) to align with the 1.5°C goal. • A work plan for enhanced climate action financing was introduced, titled "Baku to Belém Roadmap to 1.3T," guiding efforts toward meeting the new financial targets.

List of Contributors

Christophe Aura, current President at COPTICOM
COPTICOM is a public affairs firm focused on environmental protection, climate action, and reduction of social inequalities. Christophe also served for over three years as a senior advisor in government affairs and decarbonization at ArcelorMittal Mining Canada. His diverse experiences at both ArcelorMittal and COPTICOM have allowed him to contribute to projects spanning clean technologies, sustainable mobility, responsible finance, land protection and more. In recognition of his efforts, Christophe was awarded the "Impact Generation" Award in 2024 by the Junior Chamber of Commerce of Montreal, honouring his dedication to enhancing environmental and social impacts within his professional sphere.

Caroline Brouillette, Executive Director of Climate Action Network Canada
Caroline Brouillette is the executive director of Climate Action Network Canada, the country's farthest-reaching network of organizations working on climate and energy issues, with over 180 members from coast to coast to coast. She is the first francophone executive director of the network and works to create

strong and broad social consensus for climate solutions that address the convergence of crises the world is faced with, both through national policy development and in international diplomacy forums. Caroline represents environmental NGOs on Canada's Sustainable Jobs Partnership Council, a legislated body providing advice to the federal government on measures and policies related to the labour impacts of the transition. She also co-chairs the Just Transition working group at CAN International, a global network of more than 1,900 civil society organizations in over 130 countries. Caroline's commentary on climate issues has appeared widely in Canadian and international media, including in CBC Radio-Canada, *the New York Times*, Reuters, and Al Jazeera. She joined the climate movement in 2018, when she represented Canadian youth at the G7 summit in Charlevoix. Caroline holds a master's in public policy from the Lee Kuan Yew School of the National University of Singapore, where she majored in economics.

Miyuki Qiajunnguaq Daorana, MA Candidate and Inuit Youth Activist
Miyuki Qiajunnguaq Daorana is from the Inughuit society, a subgroup of Inuit from Northern West Greenland. She is a master's student in Indigenous studies, holds a bachelor's in anthropology, and has done various international and regional advocacy work for Indigenous Knowledge, Rights, and Climate Justice. She also works as an anthropological interpreter between Inuktun and English for filmmakers and researchers.

Dane de Souza, Senior Policy Advisor on Emergency Management for the Métis National Council
Dane de Souza is a proud citizen of the Métis Nation of Alberta and the senior policy advisor on emergency management for the Métis National Council. Dane began his emergency management career as a Helitack Wildland Firefighter based out of Nordegg, Alberta, for a total of six seasons. What started as a summer job to pay for university quickly turned into a lifelong

passion and path to connecting to his Métis heritage. Dane graduated in 2019 from the University of British Columbia with a master's degree in international forestry, with a focus on the impacts of climate change on wildfires, Indigenous communities, and international networks designed to aid those most impacted by climate change. Dane joined the Métis National Council's environment team in spring 2022, bringing with him contemporary insights into climate change mitigation and adaptation, as well as boots-on-the-ground experience in helping Métis communities impacted by wild-fires. Dane is currently embarking on a path to better serve the Métis community as it copes with challenges exacerbated by the impacts of climate change. Through this work, Dane hopes to further truth and reconciliation via climate adap-tation founded upon the unique connection between Métis Peoples and the land, as well as provide a voice to the concerns of Métis communities impacted by climate change.

Dalee Sambo Dorough, Iñupiaq Advocate and Lawyer
Dalee Sambo Dorough is an influential Alaskan Iñupiaq leader specializing in human rights law, Indigenous empow-erment, and Arctic governance. She holds a master's from the Fletcher School of Law and Diplomacy, Tufts University, and a PhD from the Faculty of Law at the University of British Columbia. Even before her formal education, she was active in advocating for Indigenous rights, notably receiving certifi-cation as a Tribal Court Advocate, while still in high school. Dr. Dorough's career spans numerous leadership roles, includ-ing serving as the executive director of the Inuit Circumpolar Council from 1982 to 1989 and later leading the International Union for Circumpolar Health and the Alaska Inter-Tribal Council. She spent 11 years as an associate professor at the University of Alaska Anchorage before returning to chair the Inuit Circumpolar Council from 2018 to 2022. Currently, she is a senior scholar and special advisor on Arctic Indigenous Peoples at the same university. Her work includes voluntary

leadership positions within international organizations such as the UN Permanent Forum on Indigenous Issues and involvement with the UN Framework Convention on Climate Change. She has been instrumental in advancing the UN Declaration on the Rights of Indigenous Peoples and the International Labor Organization's Convention No. 169. She presently holds the Arctic Region seat on the UN Expert Mechanism on the Rights of Indigenous Peoples, an advisory body to the Human Rights Council. Her advocacy has earned her numerous accolades, including the Reebok Human Rights Award in 1988 and the International Arctic Science Committee Medal in 2022. In 2020, she was recognized by *USA Today* as one of Alaska's ten most influential women over the past century. In June 2024, she received an honorary doctorate from the University of Durham for her work in human rights, empowerment of Indigenous Peoples, and Arctic governance.

The Honourable Rosa Galvez, Senator
Rosa Galvez, originally from Peru, is one of Canada's leading experts in pollution control and its effects on human health. She has a PhD in environmental engineering from McGill University and was a professor at Laval University in Québec for over 25 years, heading the Civil and Water Engineering Department from 2010 to 2016. She specializes in water and soil decontamination, waste management and residues, sustainable development, environmental impact assessments, and climate risk to infrastructure. Throughout her career, she has been requested by private, governmental, and community organizations to offer expert advice. She has advised a number of international organizations, including on Canada–US and Quebec–Vermont agreements regarding the protection of the Great Lakes and the St. Lawrence River. She also conducted an important study on the catastrophic oil spill at Lac-Mégantic. Senator Galvez is a member of the Ordre des ingénieurs du Québec and the Pan-American Union of Engineering Societies. She is also a Fellow of Engineers Canada, the Canadian

Society for Civil Engineering, and the Canadian Academy of Engineering. Her research has led her around the world to countries such as France, Italy, Belgium, Japan, and China. Senator Galvez was appointed to the Senate on December 6, 2016, representing Quebec (Bedford). She has served on several parliamentary committees, including the Standing Senate Committee on Energy, the Environment and Natural Resources, the Standing Senate Committee on National Finance, and the Standing Senate Committee on Transport and Communications.

Kate Gillis, Senior Policy Advisor for International Relations at the Métis National Council
Kate Gillis is the senior policy advisor for international relations at the Métis National Council. Born and raised in Calgary, Alberta, Kate is a proud citizen of the Métis Nation, with family roots in St. Laurent, Manitoba. Holding a master of arts in Indigenous Studies (University of Saskatchewan), Kate's thesis utilized Métis worldviews of *wahkohtowin* (kinship) and *Otipemisiwak* (the people who own themselves) to articulate the role of Métis women in nation building. In her current role at MNC, Kate works across departments to ensure that all international engagement takes a rights-based approach and is grounded in these key Métis principles. Prior to joining MNC, Kate was the first ever Indigenous intern at the Permanent Mission of Canada to the United Nations. She is passionate about upholding and uplifting international Indigenous solidarity networks and the fight for the enhanced participation of Indigenous Peoples within the United Nations system.

Piita Irniq, Elder and Knowledge Holder
Elder Piita Irniq (formerly known as Peter Irniq and Peter Ernerk) was born in 1947 at Lyon Inlet, Northwest Territories, today Nunavut. He is a cultural proponent, artist, public servant, and commissioner of Nunavut. Irniq represented the Keewatin region in the Council of the Northwest Territories from 1975 to 1979. From 2000 to 2005, he served as the second commissioner

of Nunavut. Irniq has worked endlessly to preserve and promote Inuit culture and languages across Canada and the world.

Richard Kinley, Former UN Assistant Secretary-General/ UNFCCC Deputy Executive Secretary
Richard Kinley was a staff member of the UNFCCC Secretariat from 1993 to 2017. During that time, he led the Secretariat team supporting the negotiations of the Kyoto Protocol (1995–1997), served as secretary of the COP (1997–2009) and deputy executive secretary (2006–2017), and managed the organization and political preparation of the Paris Conference (2015). He has advised 25 COP presidents in different capacities and was the senior secretariat expert on the international climate change process. Prior to going to the United Nations, he was an official of the Government of Canada with responsibilities relating to international environmental policy and climate change. He is currently president of the Foundation on Global Governance and Sustainability (FOGGS).

Lisa Qiluqqi Koperqualuk, former Vice-Chair, Inuit Circumpolar Council
Lisa Qiluqqi Koperqualuk was born in Puvirnituq, Northern Quebec (Nunavik). Raised by her grandparents, her elementary schooling was done in Nunavik. With a bachelor's degree in political science from Concordia University, Montréal, she holds a master's degree in anthropology from Laval University, Québec City. Fluent in Inuktitut, English, and French, Lisa acted as communications officer for Makivvik Corporation for seven years and participated in various regional, national, and international forums. Lisa served as ICC Canada's vice-president of international affairs from 2018 to 2022 before being acclaimed as president at the 14th General Assembly in July of 2022, serving as president 2022-2025. As VP of international affairs, Lisa focused much of her work in the areas of international shipping regulations, successfully leading ICC to receive provisional status at the International Marine Organization

(IMO) where Inuit are the first Indigenous people. She works for Inuit interests in self-determination, advocating for Inuit political and economic autonomy, social justice (particularly through Inuit law), and the protection of the environment, culture, and language.

Alexina Kublu
Alexina Kublu was born in Igloolik but has lived in several communities throughout Nunavut. She has made her home in Iqaluit since 1992. Kublu holds a bachelor's of education from McGill through the Nunavut Teacher Education program. She is a certified Inuktitut-English translator. She has been an instructor in the Language and Culture Program at Nunavut Arctic College; served as a member of the board for the Canadian Association for Suicide Prevention; been a board member for the Kamatsiaqtut Nunavut Help Line; served on the Federal Ministerial Task Force on Aboriginal Languages and Cultures; was chairperson for the Akitsiraq Law School Society; held the position of senior justice of the peace for the Nunavut Justices of the Peace program; taught Inuktitut online at the University of Washington. Kublu is also a former Nunavut Languages Commissioner, serving in this role from 2008 to 2012.

Susie-Ann Kudluk, Former National Inuit Youth Council President
Susie Ann Kudluk is the former president of the National Inuit Youth Council and vice-president of Qarjuit Youth Council from the Nunavik Region of Canada. Advocating for Inuit youth since the age of 23, she works towards amplifying the voices of Inuit youth across Canada, and pushing positive changes for the lives of not just Inuit youth but for all Inuit. With the help of Inuit Tapiriit Kanatami, Inuit Circumpolar Council, and the Inuit land claims organizations, Makivvik Corporation, Inuvialuit Regional Corporation, Nunatsiavut Government and Nunavut Tunngavik Incorporated, the

National Inuit Youth Council work towards creating a better and more sustainable environment for future generations.

Jean Lemire, Quebec's Climate Change Envoy
A biologist by training, Jean Lemire is an environmental specialist and renowned science communicator. He was appointed Special Envoy for Climate Change, Nordic and Arctic Affairs by the Government of Quebec in September 2017, thus becoming the first envoy in the history of Quebec diplomacy. In this capacity, he develops international partnerships and represents the Government of Quebec at various multilateral conventions and forums, including the United Nations Framework Convention on Climate Change, the Convention on Biological Diversity, the Arctic Council, the Beyond Oil and Gas Alliance and the United Nations Environment Programme's Intergovernmental Negotiating Committee on Plastic Pollution. As part of his responsibilities, Jean Lemire acts as: Representative and spokesperson for subnational governments at the Convention on Biological Diversity; Ambassador for the SAGA project on food security and climate change of the United Nations Food and Agriculture Organization; President of the Advisory Board of the International Climate Cooperation Program (ICCP), winner of the United Nations Global Climate Action Award; and, Coordinator of the Government of Quebec's International Environmental Team. His work on climate change and biodiversity has led to major international scientific missions. His films, books and web creations have garnered numerous awards. Among his distinctions, he was named an Officer of the Order of Canada, has received two honorary doctorates (oceanography-UQAR and geography-UQAM) and holds the title of Grand Ambassador of the Université de Sherbrooke. In 2010, UN Secretary-General Ban Ki Moon appointed him UN Green Wave Ambassador for biodiversity. He is also the first recipient of the prestigious Midori Prize for Biodiversity.

Elizabeth May, Member of Parliament
Elizabeth May served as leader of the Green Party of Canada from 2006 to 2019 and returned as leader in November 2022. She is the Green Party of Canada's first elected member of Parliament, representing Saanich-Gulf Islands since 2011. She now serves as the co-leader of the Green Party of Canada with Jonathan Pedneault. In 2005, Elizabeth May was made an officer of the Order of Canada in recognition of her decades of leadership in the Canadian environmental movement. She graduated from Dalhousie Law School and was admitted to the bar in both Nova Scotia and Ontario. She practised law in Ottawa with the Public Interest Advocacy Centre prior to becoming senior policy advisor to the federal minister of the environment (1986–1988). For 17 years Elizabeth served as executive director of the Sierra Club of Canada. A proud mother and grandmother, she lives in Sidney, British Columbia, with her husband, John Kidder. Elizabeth is the author of nine books, including her most recent revised book, *Climate Change for Dummies.*

David Miller, Managing Director, C40 Centre for City Climate Policy and Economy
David Miller is the managing director of the C40 Centre for City Climate Policy and Economy, the author of *Solved: How the Great Cities of the World Are Fixing the Climate Crisis* and the former mayor of Toronto. He has held a variety of public and private positions and is the recipient of honorary doctorates from York University and the University of Waterloo. A lawyer professionally, he is married to lawyer Jill Arthur. David and Jill are the parents of two children.

Sara Olsvig, Chair, Inuit Circumpolar Council
Sara Olsvig is the international chair of the Inuit Circumpolar Council. Olsvig is a long-time Indigenous Peoples' rights and human rights defender, and a politician who has served as member of the Parliament of Denmark (2011–2015) and the Parliament of Greenland (2013–2018). Sara Olsvig actively

contributed to the work of the Constitutional Commission of Greenland as well as the Human Rights Council of Greenland. Olsvig holds a master of science in anthropology and is currently a PhD candidate, Institute of Social Science, Economics & Journalism, Department of Arctic Social Science & Economics at Ilisimatusarfik, the University of Greenland. Olsvig is a recipient of the 2023 Womenomics Inclusion Award. Sara Olsvig is Inuk, and was born in Nuuk, Greenland, where she resides with her partner and their children.

Mark Purdon, Professor in the Department of Strategy, Social and Environmental Responsibility and Chair in Decarbonization at the École des sciences de la gestion, Université du Québec à Montréal

Mark Purdon is an expert in the emerging field of comparative climate change politics, which combines elements of comparative politics, comparative public policy, and international relations. He is particularly interested in the relationship between decarbonization and political economy, and has extensive research experience in both developing and developed countries in areas of climate finance, renewable energy, transportation, forestry, and land use. Other elements of Professor Purdon's research include global climate governance and international cooperation, with a particular interest in carbon markets and other climate finance instruments. His goal is to improve understanding of climate change politics—at different levels of analysis—through rich and contextualized comparative research in order to identify effective policy interventions for decarbonization. Mark earned a PhD in political science at the University of Toronto in 2013 and completed a postdoctoral fellowship at the London School of Economics in 2014. He joined the Department of Strategy, Social and Environmental Responsibility at ESG UQAM in 2018, where he has held the Chair in Decarbonization since 2021. He also leads UQAM's delegation to the annual UN climate change conference. His book *The Political Economy of Climate Finance Effectiveness in Developing Countries: Carbon Markets, Climate Funds, and the State* was published by Oxford University Press in 2024.

Graeme Reed, Strategic Advisor at the Assembly of First Nations
Graeme Reed works at the Assembly of First Nations leading their involvement in federal and international climate policy, including as outgoing Indigenous North American representative of the Facilitative Working Group on Local Communities and Indigenous Peoples Platform. He holds a PhD from the University of Guelph and is from Ottawa. He has mixed Anishinaabe, English, German, and Scottish ancestry.

Patrick Rondeau, Environmental and Just Transition Department Director, Fédération des travailleurs et des travailleuses du Québec
Patrick Rondeau is the environment and just transition department director of Fédération des travailleurs et des travailleuses du Québec (FTQ). He was FTQ regional advisor for the Montréal greater metropolitan area from 2012 to 2019. He was also in charge of mobilization and coordinated several national FTQ campaigns. He was a member of the coordinating committee of the Front commun pour la transition énergétique from 2016 to 2021. He has been a member of the Pôle d'économie circulaire since 2019. In 2018, he coordinated the Summit for a Just Energy Transition, at the Palais des congrès, in Montréal. He was a member of the electrification working group for the development of the Quebec government's Plan d'électrification et de lutte aux changements climatiques, and was a member of the Government of Canada's Advisory Council on Sustainable Development as well as an expert on the advisory committee of the Pôle d'expertise en transition verte. Since 2024, he has been a member of the board of directors of Climate Action Network Canada. He has coordinated various FTQ delegations to the UN Conferences of the Parties (COP) since 2015. Patrick

also represents the union constitution as co-focal point at the United Nations Framework Convention on Climate Change. He has also participated in various panels and given conferences as an expert on just transition since 2016. Patrick has been active in the labour movement since 2000.

Anne Simpson, ICC Climate Change Advisor
Anne Adams Simpson is of English and Scottish descent and was raised in Tk'emlúps to Secwépemc territory, situated within the unceded ancestral lands of the Secwépemc Nation of what is now Kamloops, British Columbia. Anne has a Bachelor of Arts from the University of Ottawa and a master of arts in international affairs from Carleton University, which allowed her to move briefly to the Northwest Territories for her studies. Beginning her career at Crown-Indigenous Relations and Northern Affairs Canada, Anne went on to join the Inuit Circumpolar Council in 2023 as the policy advisor on climate change and recently returned to continue her work in the public service at Indigenous Services Canada. Anne's passion for the North first came from her grandfather, Peter Adams, who was a storyteller, glaciologist, and politician. Outside of work, Anne is the vice-president of the Arctic Circle Ottawa, a local group of Ottawans who are passionate about the North, and has been a volunteer at the Canadian Museum of Nature for more than 10 years; she also loves birdwatching with friends. Anne lives in Ottawa with her partner and two cats.

Dominique Souris, Social Entrepreneur and Impact Strategist
Dominique Souris is an award-winning social entrepreneur and impact strategist working at the nexus of climate justice, technology, and capital. She is the head of climate at Goodwall, a mobile-first platform equipping 3M+ youth for the future of work on a changing planet. With nearly 15 years of experience, her background includes co-founding Youth Climate Lab, advising the Seychelles in UNFCCC negotiations, and managing a climate justice donor collaborative. Dominique

also advises climate investors, philanthropic initiatives, and think tanks on unlocking capital and scaling impact.

John Stackhouse, Senior Vice President, Office of the CEO, Royal Bank of Canada
John Stackhouse is senior vice-president in the Office of the CEO at Royal Bank of Canada, leading the organization's policy research on economic, technological, social, and climate issues. Previously, John was editor-in-chief of *The Globe and Mail* and editor of *Report on Business.* He is a senior fellow at the C. D. Howe Institute and the Munk School for Global Affairs and Public Policy, and sits on the board of the Canadian International Council. John has written four books: *Planet Canada: How Our Expats Are Shaping the Future, Mass Disruption: On the Frontlines of a Media Revolution, Timbit Nation: A Hitchhiker's View of Canada,* and *Out of Poverty.* He has contributed to four other books, including *Re-Imagining Capitalism: Building a Responsible Long-Term Model,* with Dominic Barton, Dezso Horvath, and Matthias Kipping. John speaks and writes regularly on Canada's place in the world, and hosts the popular podcast RBC Disruptors.

Catherine Stewart, Canada's Former Ambassador for Climate Change
Catherine Stewart has served in leadership roles for the Government of Canada at eight consecutive UN Framework Convention on Climate Change COPs, including as chief negotiator and Canada's Ambassador for Climate Change. She has also served in various senior executive positions at Environment and Climate Change Canada since 2014, including assistant deputy minister of international affairs. With almost 30 years in the federal government, she has in-depth policy and management experience in a wide variety of areas, including environment, climate, defence and security, military procurement, and government decision-making. Prior to joining government, Catherine worked on a United Nations

peacekeeping and electoral mission in Mozambique. She holds a bachelor of arts from McGill University and a master of arts in public administration from Carleton University.

Berry Vrbanovic

Berry was re-elected as mayor of Kitchener, Ontario, in 2022 after serving many years as a city councillor. In his role as mayor, he is also a member of the regional council. Berry was born in Zagreb, Croatia. Berry attended St. Jerome's High School and graduated from Wilfrid Laurier University with a bachelor's in political science and a diploma in business administration.

Sheila Watt-Cloutier

Sheila Watt-Cloutier is an Inuk leader from Kuujjuaq, Nunavik (northern Quebec). She is deeply committed to environmental advocacy, Indigenous rights, and youth empowerment. She has spent much of her life working to bridge Inuit Knowledge with global environmental and political negotiations. She has been a key spokesperson in global United Nations negotiations, particularly in advocating for the elimination of persistent organic pollutants that impact Arctic food systems. Her work extends to climate change advocacy, where she has brought Inuit perspectives to international human rights discussions, including filing a climate change-related petition to the Inter-American Commission on Human Rights in 2005. Her contributions have earned her global recognition, including the United Nations Champion of the Earth Award, the Right Livelihood Award (often referred to as the "Alternative Nobel Prize"), the Sophie Prize in Norway, and the Climate Change Award from the Prince Albert of Monaco Foundation. She has received 22 honorary doctorates from institutions across Canada and one from the United States, and in 2006, she was appointed an officer of the Order of Canada. She was also nominated for the Nobel Peace Prize in 2007 for her pioneering work linking climate change and human rights. Her latest recognition is a Lifetime Achievement Award from the National Geographic Society of Spain (2024).

Duo

A New Imprint Dedicated to Literary Works and Ideas

Duo presents works of fiction and non-fiction, in French and in English, as well as in Indigenous languages, from established writers and emerging talents. It is one of the imprints of Les Presses de l'Université d'Ottawa / University of Ottawa Press (PUO-UOP).

Other Literary Titles Published by the University of Ottawa Press

Works of Fiction

Abla Farhoud, *Happiness Has a Slippery Tail*, translated from the French by Judith Weisz Woodsworth, 2025.

Jean-Louis Grosmaire, *Mouvance et espérance*, 2025.

Yolande Bastarache, *Détresse et nostalgie*, 2024.

Robert Major, *Éloge de la procrastination et autres facéties*, 2022.

Michel Picard, *Kilis*, 2021.

Jean-Louis Grosmaire, *Acadissima*, 2021. Prix France-Acadie.

Yolande Bastarache, *Mon village, la côte*, 2021.

Maurice Henrie, *Odette*, 2020.

Michel Picard, *Memoriam*, 2020.

Camilla Grudova, *L'alphabet des poupées*, translated from the English by Véronique Lessard and Marc Charron, 2020.

Maurice Henrie, *La maison aux lilas*, 2019.

Andrew F. Sullivan, *Tout vient à mourir*, translated from the English by Marc-André Clément, 2017.

Literary Non-fiction

Pierre Calvé, *Le français grandeur nature : portrait et défense d'une langue vivante*, 2025.

Nicole V. Champeau, *Pointe Maligne, retrouvée par les textes : présence française dans le Haut Saint-Laurent* (tome II), 2023.

Robert Major, *Identité, appartenances : un parcours franco-ontarien*, 2023.

Maurice Henrie, *La tête haute*. 2022. Prix du livre d'Ottawa.

Robert Major, *Témiscamingue : châtiments, miracles, et autres propos du concierge de l'évêché*, 2021.

Robert Major, *Carnets du rang 5 : fragments d'un enracinement, fragments d'un parcours*, 2021.

Robert Major, *Mes conversations avec Claude*, 2019.

Biographies and Memoirs

Monique Aubry-Frize, *Une femme en ingénierie : mémoires d'une pionnière*, translated from the English by Suzanne Aubry, 2024.

Roy MacGregor, *L'étoile du nord : le mystère éternel de Tom Thomson et de la femme qui l'aimait*, translated from the English by Benoit Léger, 2023.

Fred Langan, *Elle a osé réussir : biographie de l'honorable Marie-P. Charette-Poulin / She Dared to Succeed: A Biography of the Honourable Marie-P. Charette-Poulin*, 2023.

Michel Bastarache and Antoine Trépanier, *What I Wish I Had Told My Children*, translated from the French by Julie Da Silva, 2023.

Stéphane Desjardins, *La famille Fermanian : l'histoire du cinéma Pine de Sainte-Adèle*, 2022.

Michel Bastarache and Antoine Trépanier, *Michel Bastarache : ce que je voudrais dire à mes enfants*, 2019.

Discover Duo and all works of fiction and literary non-fiction here:
www.duoeditions.ca

For a complete list of titles published
by the University of Ottawa Press, please visit:
www.Press.uOttawa.ca